# Groups

Modular Mathematics Series

# Groups

C R Jordan & D A Jordan
School of Mathematics and Statistics,
University of Sheffield

Edward Arnold
A member of the Hodder Headline Group
LONDON MELBOURNE AUCKLAND

First published in Great Britain 1994

*British Library Cataloguing in Publication Data*

Available upon request

ISBN 0 340 61045 X



Typeset in ten point Times by Page Bros (Norwich) Ltd.
Printed and bound in Great Britain for Edward Arnold, a division of Hodder Headline PLC, 338 Euston Road, London NW1 3BH by St. Edmundsbury Press Ltd, Bury St. Edmunds, Suffolk and Hartnolls Ltd, Bodmin, Cornwall.

# Series Preface

This series is designed particularly, but not exclusively, for students reading degree programmes based on semester-long modules. Each text will cover the essential core of an area of mathematics and lay the foundation for futher study in that area. Some texts may include more material than can be comfortably covered in a single module, the intention there being that the topics to be studied can be selected to meet the needs of the student. Historical contexts, real life situations, and linkages with other areas of mathematics and more advanced topics are included. Traditional worked examples and exercises are augmented by more open-ended exercises and tutorial problems suitable for group work or self-study. Where appropriate, the use of computer packages is encouraged. The first level texts assume only the A-level core curriculum.

Professor Chris D. Collinson
Dr Johnston Anderson
Mr Peter Holmes

# Preface

Groups are sets of objects, which might be numbers, matrices or functions, in which objects can be combined, subject to a list of basic rules or axioms. The general theory is developed logically from these axioms. Although this idea of axiomatization is important, it can make the subject appear unduly abstract. Therefore we have made very effort, throughout this book, to illustrate new ideas through concrete examples.

Groups arise most naturally in considerations of symmetry. In particular, the symmetry of a geometrical figure, like a square or circle, can be measured by its group of symmetries. These groups are particularly helpful in illustrating many basic notions. The first chapter informally discusses these groups and introduces many ideas which are formalized in the book.

We introduce the idea of a group acting on a set earlier than is perhaps usual. This involves thinking of elements of groups as doing something to elements of another set, rather than as things satisfying a seemingly arbitrary list of axioms. for example, in the group of symmetries of the square, the elements of the group rotate or reflect the points of the square.

There are numerous exercises in the book. Those which are contained within chapters are usually short. Some are simply intended to help in understanding a new concept. Others encourage investigation of examples and, occasionally, prediction of results to be proved later. Solutions to these exercises are provided at the end of the book.

Each chapter ends with a section of further exercises. Some of these contain the word 'investigate' as an indication that the exercise is intended to be open ended. Many of these exercises would be suitable for tutorial discussion or workshop projects. Some suggestions for further projects, investigations and reading are included in the last chapter. This chapter also contains some hisotircal notes on the development of the subject. These notes include brief comments on several mathematicians whose names appear in bold type elsewhere in the book.

The book is intended for use on a twenty-four lecture modular course. As there is variety in the timing of the introduction of group theory in different degree courses, and consequently in the background knowledge of students, we have included more material then we would expect to cover in twenty-four lectures. The first twelve chapters provide one possible course, stopping short of factor groups. The book could also be used by anyone who wishes to teach group theory as far as factor groups to students already familiar with gneeral ideas such as functions and equivalence relations.

There are many people to whom thanks are due. Over the years we have benefited from the cooperation and ideas of out colleagues in Sheffield. We would particularly like to thank John Greenlees, who made several suggestions, and the series editor Johnston Anderson, who made a number of comments. On a more practical note we should thank Mike Piff who helped us with LaTeX , in which we types the original manuscript, and with the *mpfic* macros which we used for the pictures. We should also thank our sons Jonathan and Thomas who have tolerated many a mealtime discussion of 'the book'. Finally we would like to thank the staff of Arnolds, both past and present, for their encouragement and help.

Camilla Jordan
David Jordan
*February 1994*

# Contents


Series Preface . . . v

Preface . . . vii

1 **Squares and Circles** . . . 1
- 1.1 Symmetries of a square . . . 1
- 1.2 Symmetries of a circle . . . 7
- 1.3 Further exercises on Chapter 1 . . . 9

2 **Permutations** . . . 12
- 2.1 The symmetric group $S_4$ . . . 12
- 2.2 Functions . . . 14
- 2.3 Permutations . . . 19
- 2.4 Basic properties of cycles . . . 21
- 2.5 Cycle decomposition . . . 22
- 2.6 Transpositions . . . 24
- 2.7 The 15-puzzle . . . 26
- 2.8 Further exercises on Chapter 2 . . . 27

3 **Linear Transformations and Matrices** . . . 29
- 3.1 Matrix multiplication . . . 29
- 3.2 Linear transformations . . . 32
- 3.3 Orthogonal matrices . . . 34
- 3.4 Further exercises on Chapter 3 . . . 36

4 **The Group Axioms** . . . 38
- 4.1 Number systems . . . 38
- 4.2 Binary operations . . . 40
- 4.3 Definition of a group . . . 41
- 4.4 Examples of groups . . . 41
- 4.5 Consequences of the axioms . . . 43
- 4.6 Direct products . . . 46
- 4.7 Further exercises on Chapter 4 . . . 47

5 **Subgroups** . . . 49
- 5.1 Subgroups . . . 49
- 5.2 Examples of subgroups . . . 50
- 5.3 Groups of symmetries . . . 52
- 5.4 Further exercises on Chapter 5 . . . 55

6 **Cyclic Groups** . . . 57
- 6.1 Cyclic groups . . . 57
- 6.2 Cyclic subgroups . . . 58
- 6.3 Order of elements . . . 59
- 6.4 Orders of products . . . 61
- 6.5 Orders of powers . . . 62

| | | | |
|---|---|---|---|
| | 6.6 | Subgroups of cyclic groups | 63 |
| | 6.7 | Direct products of cyclic groups | 65 |
| | 6.8 | Further exercises on Chapter 6 | 65 |
| **7** | **Group Actions** | | **68** |
| | 7.1 | Groups acting on sets | 68 |
| | 7.2 | Orbits | 71 |
| | 7.3 | Stabilizers | 72 |
| | 7.4 | Permutations arising from group actions | 73 |
| | 7.5 | The alternating group | 74 |
| | 7.6 | Further exercises on Chapter 7 | 78 |
| **8** | **Equivalence Relations and Modular Arithmetic** | | **81** |
| | 8.1 | Partitions | 81 |
| | 8.2 | Relations | 82 |
| | 8.3 | Equivalence classes | 84 |
| | 8.4 | Equivalence relations from group actions | 85 |
| | 8.5 | Modular arithmetic | 87 |
| | 8.6 | Further exercises on Chapter 8 | 91 |
| **9** | **Homomorphisms and Isomorphisms** | | **93** |
| | 9.1 | Comparing $D_3$ and $S_3$ | 93 |
| | 9.2 | Properties of homomorphisms | 97 |
| | 9.3 | Homomorphisms arising from group actions | 98 |
| | 9.4 | Cayley's theorem | 102 |
| | 9.5 | Cyclic groups | 103 |
| | 9.6 | Further exercises on Chapter 9 | 104 |
| **10** | **Cosets and Lagrange's Theorem** | | **106** |
| | 10.1 | Left cosets | 106 |
| | 10.2 | Left cosets as equivalence classes | 108 |
| | 10.3 | Lagrange's theorem | 110 |
| | 10.4 | Consequences of Lagrange's theorem | 111 |
| | 10.5 | Applications to number theory | 112 |
| | 10.6 | Right cosets | 114 |
| | 10.7 | Further exercises on Chapter 10 | 114 |
| **11** | **The Orbit-Stabilizer Theorem** | | **117** |
| | 11.1 | The orbit-stabilizer theorem | 117 |
| | 11.2 | Fixed subsets | 119 |
| | 11.3 | Counting orbits | 121 |
| | 11.4 | Further exercises on Chapter 11 | 122 |
| **12** | **Colouring Problems** | | **124** |
| | 12.1 | Colouring problems | 124 |
| | 12.2 | Groups of symmetries in three dimensions | 127 |
| | 12.3 | Three-dimensional colouring problems | 129 |
| | 12.4 | Further exercises on Chapter 12 | 130 |
| **13** | **Conjugates, Centralizers and Centres** | | **133** |
| | 13.1 | Conjugates | 133 |
| | 13.2 | Conjugacy classes | 133 |

13.3 Conjugacy classes in $S_n$ . . . 135
13.4 Centralizers . . . 137
13.5 Centres . . . 137
13.6 Conjugates and centralizers . . . 138
13.7 Further exercises on Chapter 13 . . . 140

**14 Towards Classification** . . . 142
14.1 An action of $S_3$ on three-dimensional space . . . 142
14.2 Cauchy's theorem . . . 143
14.3 Direct products . . . 145
14.4 Further exercises on Chapter 14 . . . 148

**15 Kernels and Normal Subgroups** . . . 150
15.1 Kernels of homomorphisms . . . 150
15.2 Kernels of actions . . . 151
15.3 Conjugates of a subgroup . . . 152
15.4 Normal subgroups . . . 154
15.5 Normal subgroups and conjugacy classes . . . 157
15.6 Simple groups . . . 159
15.7 Further exercises on Chapter 15 . . . 160

**16 Factor Groups** . . . 163
16.1 Cosets of the kernel of an action . . . 163
16.2 Factor groups . . . 166
16.3 Calculations in factor groups . . . 168
16.4 The first isomorphism theorem . . . 170
16.5 Groups of order $p^2$ are Abelian . . . 172
16.6 Further exercises on Chapter 16 . . . 174

**17 Groups of Small Order** . . . 176
17.1 Groups of order $p^2$ . . . 176
17.2 Groups of order $2p$ . . . 176
17.3 Groups of order 8 . . . 178
17.4 Groups of order 12 . . . 181
17.5 Further exercises on Chapter 17 . . . 184

**18 Past and Future** . . . 186
18.1 History . . . 186
18.2 Topics for further study . . . 187
18.3 Projects . . . 190

**Solutions** . . . 193

**Glossary** . . . 203

**Bibliography** . . . 204

**Index** . . . 205

# 1 • Squares and Circles

In this chapter, we shall discuss two particular groups—the groups of symmetries of the square and the circle. These groups illustrate many of the basic concepts of group theory. Several of these will be informally introduced in this chapter and treated more formally later.

## 1.1 Symmetries of a Square

In order to calculate with symmetries easily it is helpful to use polar coordinates. Any point $(x, y)$ in the plane can be determined by its distance, $r$, from the origin and the angle, $\theta$, made with the $x$-axis by the line joining $(x, y)$ to the origin. The angle $\theta$ is measured in radians and in an anticlockwise direction. We call $r$ and $\theta$ the **polar coordinates** of the point $(x, y)$. They are related to the Cartesian coordinates $x$ and $y$ by the equations:

$$x = r\cos\theta, \quad y = r\sin\theta.$$

For example, the point $(1, 1)$ has polar coordinates $r = \sqrt{2}$ and $\theta = \frac{\pi}{4}$.

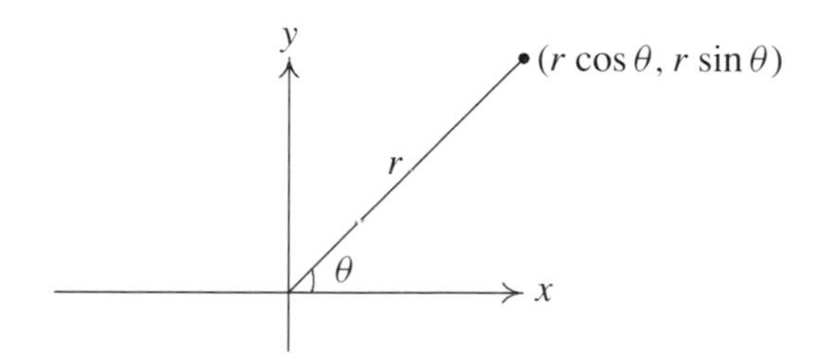


**Fig 1.1** Polar coordinates.

Consider a square in the coordinate plane with its centre at the origin and with sides parallel to the $x$ and $y$ axes. There are eight transformations of the plane which, although they may move the individual points in the square, leave the square occupying its original position. These transformations, four of which are rotations and four of which are reflections, are called the **symmetries** of the square. They are as in Fig 1.2, where rotations are anticlockwise about the origin.

These eight symmetries form a **group** which is called the **group of symmetries of the square.** It is one of an infinite family of so-called **dihedral** groups and is denoted by $D_4$. In general, $D_n$ is the group of symmetries of a regular $n$-sided polygon. In some sense $D_4$ measures the symmetry of the square. It is not just the listing of the eight elements that is significant but the fact that they can be combined together. If $f$ and $g$ are any two of these symmetries then there is a symmetry, which we denote $fg$, that has the same effect as the combined effect of first applying $g$ then $f$. This is an example of composition of functions which you may have met in calculus courses. Given a point $P = (r\cos\theta, r\sin\theta)$ in the plane, $fg(P) = f(g(P))$. Thus, to find the effect of $fg$ you apply $g$ then $f$. Note that

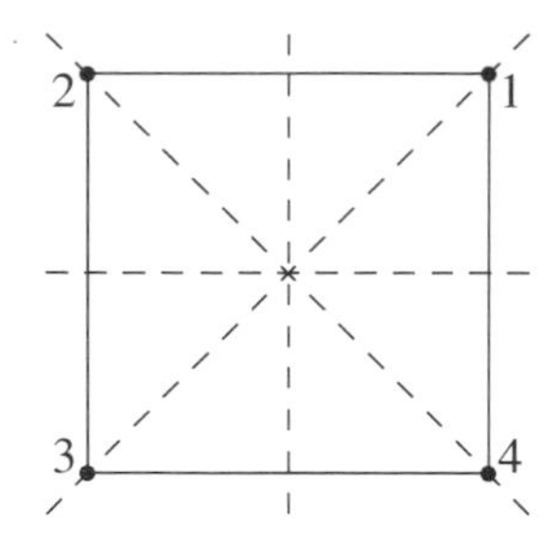


| $r_0$ | rotation through 0. |
|---|---|
| $r_1$ | rotation through $\pi/2$. |
| $r_2$ | rotation through $\pi$. |
| $r_3$ | rotation through $3\pi/2$. |
| $s_1$ | reflection in the $x$-axis. |
| $s_2$ | reflection in the $y = x$ diagonal. |
| $s_3$ | reflection in the $y$-axis. |
| $s_4$ | reflection in the $y = -x$ diagonal. |

**Fig 1.2** Symmetries of the square.

although $f$ is written first, $g$ appears next to $P$ and is applied to $P$ first. For example, consider $r_1 s_3$. To see how the composition affects the square, label the vertices 1,2,3 and 4 as shown. Then apply the transformations in turn, remembering to apply $s_3$ first, then $r_1$. You should find that 1 ends up at the bottom left and 2 ends up where it started, at top left. The overall effect is the same as that achieved by $s_4$, in other words, $r_1 s_3 = s_4$.

## EXERCISE 1

Find $s_3 r_1$, $r_1 r_2$ and $s_1 s_2$.

All 64 possible combinations $fg$ can be found in this way. However, it will be helpful, in later examples, to have formulae for the compositions of rotations and reflections. We shall denote the rotation anticlockwise through $\phi$ by $\text{rot}_\phi$. Then

$$\text{rot}_\phi(P) = \text{rot}_\phi((r\cos\theta, r\sin\theta)) = (r\cos(\theta+\phi), r\sin(\theta+\phi)).$$

This can be seen from the first diagram below where, for brevity, we have put polar coordinates rather than Cartesian coordinates in brackets. Thus, we have written $(r, \theta)$ instead of $(r\cos\theta, r\sin\theta)$.

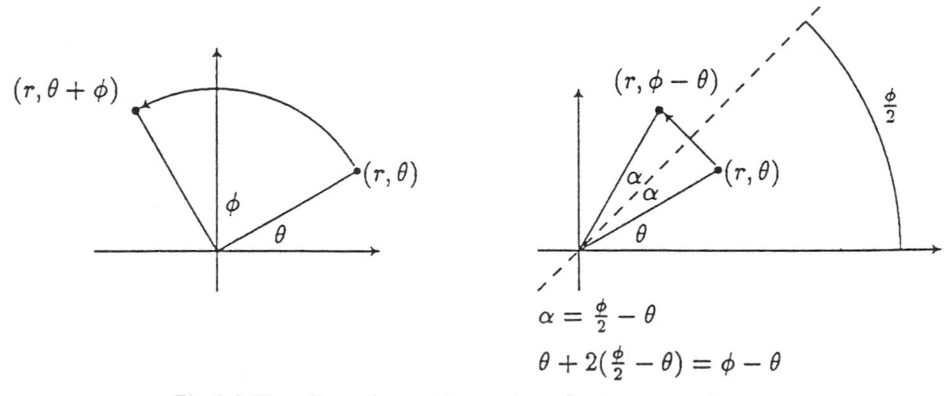


**Fig 1.3** The effect of a rotation and a reflection on a point.

The second diagram shows that if we denote the reflection in the line making an angle of $\frac{\phi}{2}$ with the $x$-axis by $\text{ref}_\phi$ then

$$\text{ref}_\phi(P) = \text{ref}_\phi((r\cos\theta, r\sin\theta)) = (r\cos(\phi-\theta), r\sin(\phi-\theta)).$$

You may be wondering why we choose $\frac{\phi}{2}$ rather than $\phi$ for the angle here. The reason is that it matches up better with the notation for rotations.

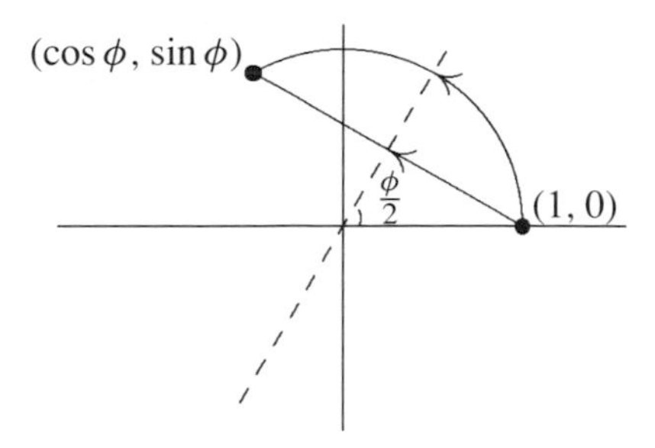


**Fig 1.4** $\text{ref}_\phi(1,0) = (\cos\phi, \sin\phi) = \text{rot}_\phi(1,0)$.

For example, notice that both $\text{rot}_\phi$ and $\text{ref}_\phi$ send the point $(1,0)$ to $(\cos\phi, \sin\phi)$. Also, the formulae which we obtain below for compositions of rotations and reflections are much neater with this choice of angle. The first of these formulae is obvious; the combined effect of rotation through $\alpha$ and $\beta$ is rotation through $\alpha+\beta$. Perhaps less obvious is the combined effect of two reflections.

$$\begin{aligned}
\text{ref}_\alpha\, \text{ref}_\beta((r\cos\theta, r\sin\theta)) &= \text{ref}_\alpha(\text{ref}_\beta((r\cos\theta, r\sin\theta))) \\
&= \text{ref}_\alpha((r\cos(\beta-\theta), r\sin(\beta-\theta))) \\
&= (r\cos(\alpha-(\beta-\theta)), r\sin(\alpha-(\beta-\theta))) \\
&= (r\cos((\alpha-\beta)+\theta), r\sin((\alpha-\beta)+\theta))
\end{aligned}$$

From this we can see that $\text{ref}_\alpha\, \text{ref}_\beta = \text{rot}_{\alpha-\beta}$, that is, the combined effect of two reflections is a rotation. Next, let us consider the combination $\text{rot}_\alpha\, \text{ref}_\beta$ of a rotation and a reflection.

$$\begin{aligned}
\text{rot}_\alpha\, \text{ref}_\beta((r\cos\theta, r\sin\theta)) &= \text{rot}_\alpha(\text{ref}_\beta((r\cos\theta, r\sin\theta))) \\
&= \text{rot}_\alpha((r\cos(\beta-\theta), r\sin(\beta-\theta))) \\
&= (r\cos((\alpha+\beta)-\theta), r\sin((\alpha+\beta)-\theta))
\end{aligned}$$

From this we can see that $\text{rot}_\alpha\, \text{ref}_\beta = \text{ref}_{\alpha+\beta}$, that is, a reflection followed by a rotation is the same as another reflection.

## EXERCISE 2

Show that $\text{ref}_\alpha\, \text{rot}_\beta = \text{ref}_{\alpha-\beta}$.

Note that this, together with the previous result, shows that the order in which the operations $\text{ref}_\alpha$ and $\text{rot}_\beta$ are applied is important. Table 1.1 summarizes the effect of combining rotations and reflections.

| |
|---|
| $\text{rot}_\alpha\, \text{rot}_\beta = \text{rot}_{\alpha+\beta}$ |
| $\text{ref}_\alpha\, \text{ref}_\beta = \text{rot}_{\alpha-\beta}$ |
| $\text{rot}_\alpha\, \text{ref}_\beta = \text{ref}_{\alpha+\beta}$ |
| $\text{ref}_\alpha\, \text{rot}_\beta = \text{ref}_{\alpha-\beta}$ |

**Table 1.1** Combining rotations and reflections.

When applying the above formulae, we sometimes obtain angles greater than $2\pi$ or less than 0. Clearly, rotations through angles that differ by $2\pi$ have the same effect, so we add or subtract $2\pi$ as appropriate. For example, $\text{rot}_{\frac{3\pi}{2}} \text{rot}_{\frac{3\pi}{2}} = \text{rot}_{3\pi} = \text{rot}_{\pi}$. The same is true for reflections because if $\phi$ and $\psi$ differ by $2\pi$ then the lines at angles $\frac{\phi}{2}$ and $\frac{\psi}{2}$ to the $x$-axis are the same. (See Fig 1.5.) For example $\text{rot}_{\frac{3\pi}{2}} \text{ref}_{\frac{3\pi}{2}} = \text{ref}_{3\pi} = \text{ref}_{\pi}$ and $\text{ref}_{\frac{\pi}{2}} \text{rot}_{\pi} = \text{ref}_{-\frac{\pi}{2}} = \text{ref}_{\frac{3\pi}{2}}$.

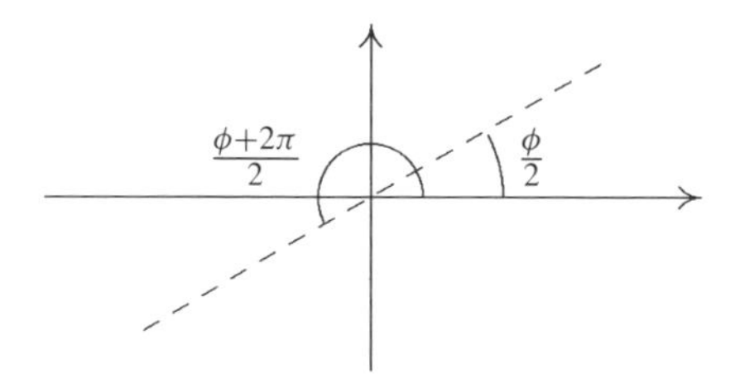


**Fig 1.5** Line of reflection for $\frac{\phi}{2}$ and for $\frac{\phi+2\pi}{2}$.

Notice that, in Table 1.1, the order matters in all but the first formula and that we have applied the right-hand operation first.

We can now complete a so-called **Cayley table** showing the combinations in $D_4$. We rewrite $r_0$ as $e$ so as to fit in with the notation of general group theory later. We use Table 1.1 with

$$r_1 = \text{rot}_{\frac{\pi}{2}},\ r_2 = \text{rot}_{\pi},\ r_3 = \text{rot}_{\frac{3\pi}{2}},\ s_1 = \text{ref}_0,\ s_2 = \text{ref}_{\frac{\pi}{2}},\ s_3 = \text{ref}_{\pi},\ s_4 = \text{ref}_{\frac{3\pi}{2}}.$$

| $\mathbf{D_4}$ | **e** | $\mathbf{r_1}$ | $\mathbf{r_2}$ | $\mathbf{r_3}$ | $\mathbf{s_1}$ | $\mathbf{s_2}$ | $\mathbf{s_3}$ | $\mathbf{s_4}$ |
|---|---|---|---|---|---|---|---|---|
| **e** | $e$ | $r_1$ | $r_2$ | $r_3$ | $s_1$ | $s_2$ | $s_3$ | $s_4$ |
| $\mathbf{r_1}$ | $r_1$ | $r_2$ | $r_3$ | $e$ | $s_2$ | $s_3$ | $s_4$ | $s_1$ |
| $\mathbf{r_2}$ | $r_2$ | $r_3$ | $e$ | $r_1$ | $s_3$ | $s_4$ | $s_1$ | $s_2$ |
| $\mathbf{r_3}$ | $r_3$ | $e$ | $r_1$ | $r_2$ | $s_4$ | $s_1$ | $s_2$ | $s_3$ |
| $\mathbf{s_1}$ | $s_1$ | $s_4$ | $s_3$ | $s_2$ | $e$ | $r_3$ | $r_2$ | $r_1$ |
| $\mathbf{s_2}$ | $s_2$ | $s_1$ | $s_4$ | $s_3$ | $r_1$ | $e$ | $r_3$ | $r_2$ |
| $\mathbf{s_3}$ | $s_3$ | $s_2$ | $s_1$ | $s_4$ | $r_2$ | $r_1$ | $e$ | $r_3$ |
| $\mathbf{s_4}$ | $s_4$ | $s_3$ | $s_2$ | $s_1$ | $r_3$ | $r_2$ | $r_1$ | $e$ |

**Table 1.2** Cayley table for $D_4$.

The composition $fg$ is entered in the intersection of the row labelled $f$ and the column labelled $g$. Several features of this table should be noted. Firstly, no new transformations are introduced by calculating compositions. In other words, whenever $f$ and $g$ are symmetries of the square, so is $fg$. We refer to this property as **closure**. Secondly, the element $e$ has the property that $ef = f = fe$ for all $f \in D_4$. This element is an example of a **neutral** or **identity** element. Every group has such an element. Thirdly, for each element $f \in D_4$ there is an element $g \in D_4$, called the **inverse** of $f$, such that $fg = e = gf$. Note that six elements of $D_4$ are their own inverses. They satisfy $f^2 = e$. The remaining two elements satisfy $f^4 = e$. Notice that $r_2 s_1 = s_1 r_2$, but $r_1 s_1 \neq s_1 r_1$. We say that $r_2$ and $s_1$ **commute** but that $r_1$ and $s_1$ do not. The existence of two elements which do not commute means that the group $D_4$ is not **abelian**. The word 'abelian' honours **N. H. Abel** and an abelian group is one in which every two elements commute.

## EXERCISE 3

Use the Cayley table (Table 1.2) to find $r_1 s_1$, $(r_1 s_1) s_3$, $s_1 s_3$ and $r_1(s_1 s_3)$. Is $(r_1 s_1) s_3 = r_1(s_1 s_3)$?

It is true that $(fg)h = f(gh)$ for all $f, g, h \in D_4$. It would be extremely tedious to check this **associative** law for all $8^3$ choices of $f, g, h$ but we shall later see that associativity always holds for composition of functions. In the main body of the table, i.e. ignoring the labelling row and column, each element of the group appears exactly once in each row and exactly once in each column. This is referred to as the **Latin square** property of Cayley tables. Of the above properties, four—namely closure, associativity, existence of a neutral element, and existence of inverses—will be built into the definition of a group. For the present we will refer to them as the group properties. We shall see later that others, such as the Latin square property, can be deduced from these.

Now number the vertices of the square as shown in Fig 1.2. Each symmetry of the square performs a **permutation** or rearrangement of the vertices. For example, $s_1$ sends 1 to 4, 2 to 3, 3 to 2 and 4 to 1. In the standard notation for permutations, this is written

$$\begin{pmatrix} 1 & 2 & 3 & 4 \\ 4 & 3 & 2 & 1 \end{pmatrix}.$$

Below each of the numbers in the top row is written the number to which it is sent by the permutation. The permutations corresponding to the eight symmetries are as follows:

$$e: \begin{pmatrix} 1 & 2 & 3 & 4 \\ 1 & 2 & 3 & 4 \end{pmatrix} \quad r_1: \begin{pmatrix} 1 & 2 & 3 & 4 \\ 2 & 3 & 4 & 1 \end{pmatrix}$$

$$r_2: \begin{pmatrix} 1 & 2 & 3 & 4 \\ 3 & 4 & 1 & 2 \end{pmatrix} \quad r_3: \begin{pmatrix} 1 & 2 & 3 & 4 \\ 4 & 1 & 2 & 3 \end{pmatrix}$$

$$s_1: \begin{pmatrix} 1 & 2 & 3 & 4 \\ 4 & 3 & 2 & 1 \end{pmatrix} \quad s_2: \begin{pmatrix} 1 & 2 & 3 & 4 \\ 1 & 4 & 3 & 2 \end{pmatrix}$$

$$s_3: \begin{pmatrix} 1 & 2 & 3 & 4 \\ 2 & 1 & 4 & 3 \end{pmatrix} \quad s_4: \begin{pmatrix} 1 & 2 & 3 & 4 \\ 3 & 2 & 1 & 4 \end{pmatrix}.$$

This situation, where each element of the group permutes the vertices, is an example of a group **acting** on a set. Note that there are two elements of $D_4$, $e$ and $s_2$ which send 1 to itself. These two elements on their own form a group with a Cayley table as in Table 1.3.

This is an example of a **subgroup**. This particular subgroup is the **stabilizer** of 1, being the set of those elements of $D_4$ which send 1 to itself. The elements sending 1 to 2 are $r_1$ and $s_3$, those sending 1 to 3 are $r_2$ and $s_4$ and those sending 1 to 4 are $r_3$ and $s_1$. Each of

| | **e** | **$s_2$** |
|---|---|---|
| **e** | $e$ | $s_2$ |
| **$s_2$** | $s_2$ | $e$ |

**Table 1.3** Cayley table for the stabilizer of 1.

these pairs of elements is an example of what we shall later call a **coset** of the above subgroup. The fact that each coset has the same number of elements as the subgroup is an instance of an important general result which we shall meet in due course.

## EXERCISE 4

Find the stabilizer of 2 and check that for each of the other vertices $V = 1, 3$ or 4, the number of elements of $D_4$ which send 2 to $V$ is the same as the number in the stabilizer.

Given any two vertices $a$, $b$, there is always an element of $D_4$ which sends $a$ to $b$. This is not true of all group actions. For example, divide the square into nine smaller squares and number these as shown in Fig 1.6.

| | | |
|---|---|---|
| 1 | 2 | 3 |
| 4 | 5 | 6 |
| 7 | 8 | 9 |

**Fig 1.6** Square divided into nine squares.

The group $D_4$ acts on these; for example the reflection $s_1$ performs the permutation

$$\begin{pmatrix} 1 & 2 & 3 & 4 & 5 & 6 & 7 & 8 & 9 \\ 7 & 8 & 9 & 4 & 5 & 6 & 1 & 2 & 3 \end{pmatrix}.$$

For a particular small square, its **orbit** consists of those squares which can be obtained by applying elements of $D_4$. For example $\{1, 3, 9, 7\}$ is the orbit of 1 and also of 3, 9 and 7.

The group $D_4$ also acts on the diagonals of the square. Number these as shown in Fig 1.7. The elements $e, r_2, s_2$ and $s_4$ send diagonal 1 to itself and diagonal 2 to itself, that is they perform the permutation $\begin{pmatrix} 1 & 2 \\ 1 & 2 \end{pmatrix}$ on the diagonals. They form the stabilizer of 1 (and also that of 2) for this action. The other four symmetries, $r_1, r_3, s_1$ and $s_3$, perform the permutation $\begin{pmatrix} 1 & 2 \\ 2 & 1 \end{pmatrix}$ on the diagonals.

## EXERCISE 5

This question refers to the action of $D_4$ on the nine small squares in Fig 1.6.

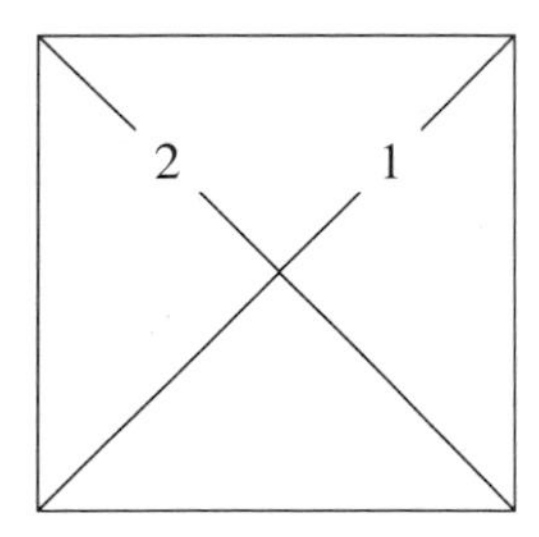


**Fig 1.7** Diagonals of a square.

(i) Find all elements of the stabilizer of (a) 2 ; (b) 5.

(ii) Find all elements of the orbit of (a) 2 ; (b) 5.

(iii) For each of 2 and 5 write down the product of the number of elements in its orbit and the number of elements in its stabilizer.

(iv) Suggest a rule for the product of the numbers of elements in an orbit and the number of elements in the corresponding stabilizer for an action of $D_4$. Does your rule work for the orbits of vertices and diagonals described in the text?

## 1.2 Symmetries of a Circle

In this section we consider the unit circle, that is the circle of radius one with its centre at the origin. A typical point on the unit circle has coordinates $(\cos\theta, \sin\theta)$ for some $\theta$.

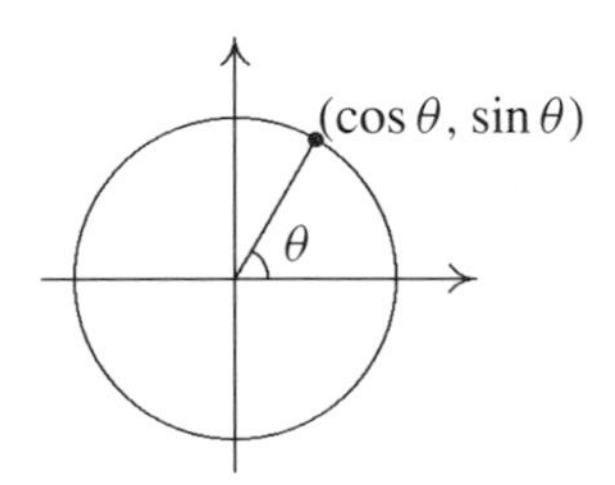


**Fig 1.8** The unit circle.

In contrast to the square, this circle has infinitely many symmetries. They are all the rotations $\text{rot}_\theta$ together with all the reflections $\text{ref}_\theta$. As for the square, these symmetries form a group, the group of symmetries of the circle, but this time the group is infinite. Clearly we cannot draw up a Cayley table as we did for the square. We need to check the group properties (closure, associativity, existence of a neutral or identity element and existence of inverses) without this. We shall not check associativity here because we shall see in Chapter 2 that it always holds for composition. We check the other group properties using the formulae given in Table 1.1. The combination of any two symmetries is either a rotation or reflection. This gives us the **closure** property. The neutral element is $rot_0$. Each rotation $rot_\theta$ has the rotation $rot_{-\theta}$ as inverse while each reflection $ref_\theta$ is its own inverse. (What is the overall effect of reflecting twice in the same line?) This group is known as the **orthogonal** group and is denoted by $O_2$. As with $D_4$ in the previous section, it is one of an infinite family of groups but this time the subscript two tells us that we are considering symmetries of a two-dimensional object.

The group $O_2$ acts on the circle by rotations and reflections. What are the orbits of this action? Consider the point $(1, 0)$. It can be sent to any point on the circumference by an appropriate rotation (or by an appropriate reflection). It cannot be sent to any point not on the circle. Thus, the orbit of $(1, 0)$ is just the circle. Indeed the orbit of any point on the circle is the circle, so there is only one orbit, namely the circle itself. Now let us consider the stabilizer of an arbitrary point $P = (\cos\theta, \sin\theta)$ on the circle. Which elements of the group $O_2$ send $P$ to itself? There are just two, namely $rot_0$ and $ref_{2\theta}$. These form the stabilizer of $P$.

## EXERCISE 6

(i) Which elements of $O_2$ stabilize the point $(0, 1)$?

(ii) Which elements of $O_2$ send the point $(0, 1)$ to the point $(-1, 0)$?

(iii) How many elements of $O_2$ send the point $(\cos\theta, \sin\theta)$ to the point $(\cos\phi, \sin\phi)$? What are they?

Now consider a disc of radius one, by which we mean the circle of radius one together with its interior. It has the same symmetries as the circle. The origin is not moved by any element of the group so it forms its own orbit and its stabilizer is the whole orthogonal group, $O_2$. Now consider any other point in the disc, say $P = (r\cos\theta, r\sin\theta)$, $1 \geqslant r > 0$. A rotation $rot_\phi$ sends $P$ to $(r\cos(\theta+\phi), r\sin(\theta+\phi))$ while a reflection $ref_\phi$ sends $P$ to $(r\cos(\phi-\theta), r\sin(\phi-\theta))$. Both these points are on the circle with radius $r$ and centre at the origin. Furthermore, if $Q = (r\cos\alpha, r\sin\alpha)$ is any other point on this circle, then $P$ is sent to $Q$ by both $rot_{\alpha-\theta}$ and $ref_{\alpha+\theta}$. Thus, the orbit of $P$ is the circle through $P$ with its centre at the origin. The only rotation to fix $P$ is $rot_0$ and the only

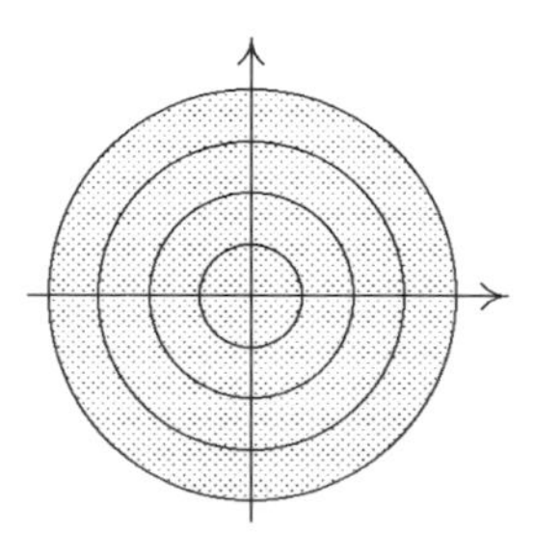

**Fig 1.9** Orbits of the action of $O_2$ on a disc.

reflection to fix $P$ is $ref_{2\theta}$ so the stabilizer of $P$ is the set $\{rot_0, ref_{2\theta}\}$.

## 1.3 Further Exercises on Chapter 1

### EXERCISE 7

Divide the square into eight isosceles right-angled triangles and number these as shown in Fig 1.10.

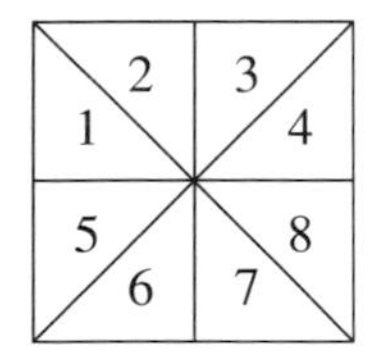


**Fig 1.10** Square divided into triangles.

(i) For each element of $D_4$, write down the corresponding permutation of $\{1, 2, 3, 4, 5, 6, 7, 8\}$.

(ii) Find all elements of the stabilizer of 1.

(iii) Find all elements of the orbit of 1.

### EXERCISE 8

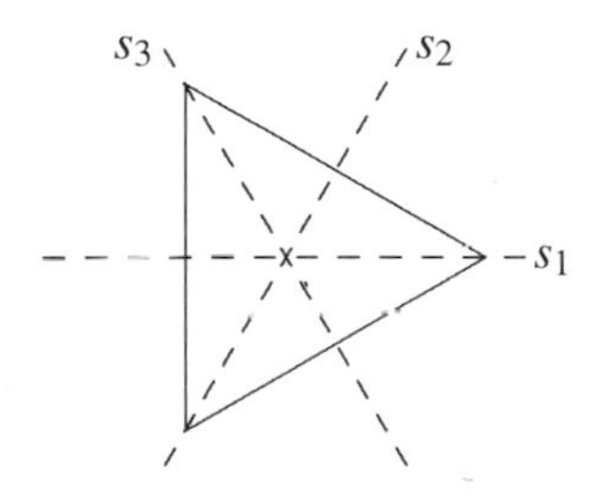


**Fig 1.11** Lines of reflection of the equilateral triangle.

The equilateral triangle shown in Fig 1.11 has six symmetries which form a group $D_3$. Complete the Cayley table shown in Table 1.4 for this group (rotations are anticlockwise).

| |
|---|
| $e = \text{rot}_0$ |
| $r_1 = \text{rot}_{\frac{2\pi}{3}}$ |
| $r_2 = \text{rot}_{\frac{4\pi}{3}}$ |
| $s_1 = \text{ref}_0$ |
| $s_2 = \text{ref}_{\frac{2\pi}{3}}$ |
| $s_3 = \text{ref}_{\frac{4\pi}{3}}$ |

| $\mathbf{D_3}$ | **e** | $\mathbf{r_1}$ | $\mathbf{r_2}$ | $\mathbf{s_1}$ | $\mathbf{s_2}$ | $\mathbf{s_3}$ |
|---|---|---|---|---|---|---|
| **e** | | | | | | |
| $\mathbf{r_1}$ | | | | | | |
| $\mathbf{r_2}$ | | | | | | |
| $\mathbf{s_1}$ | | | | | | |
| $\mathbf{s_2}$ | | | | | | |
| $\mathbf{s_3}$ | | | | | | |

**Table 1.4** Cayley table for $D_3$.

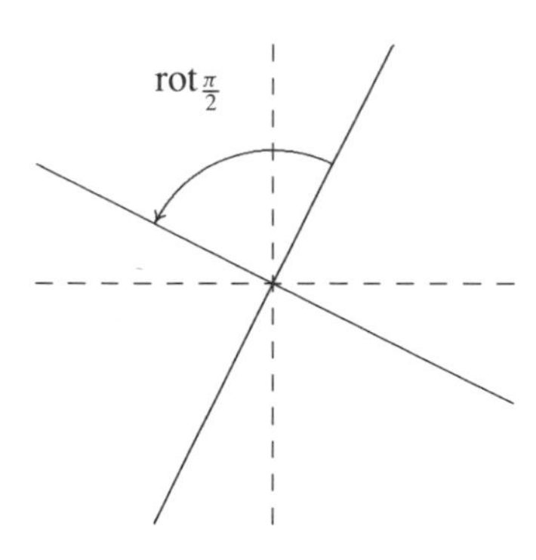


**Fig 1.12** Lines through the origin.

## EXERCISE 9

The orthogonal group $O_2$ acts on the set of all straight lines through the origin by rotation and reflection. Thus, $\mathrm{rot}_{\frac{\pi}{2}}$ sends the line $y = 2x$ to the line $y = \frac{-1}{2}x$, see Fig 1.12

(i) To which line is the line $y = x$ sent by each of the following:

$$\mathrm{rot}_{\frac{\pi}{2}},\ \mathrm{rot}_{\frac{\pi}{4}},\ \mathrm{ref}_{\pi},\ \mathrm{ref}_{\frac{3\pi}{2}}?$$

(ii) What is the orbit of the line $y = x$?

(iii) Find the four elements of $O_2$ that stabilize the line $y = x$. Draw up a Cayley table of these. (They form a subgroup of $O_2$.)

(iv) How many elements send the line $y = x$ to the line $y = -x$?

(v) To which line is the line $y = 0$ sent by each of the following:

$$\mathrm{rot}_{\frac{\pi}{2}},\ \mathrm{rot}_{\frac{\pi}{4}},\ \mathrm{ref}_{\pi},\ \mathrm{ref}_{\frac{3\pi}{2}}?$$

(vi) What is the orbit of the line $y = 0$?

(vii) Find the four elements of $O_2$ that stabilize the line $y = 0$. Draw up a Cayley table of these. (They form a subgroup of $O_2$.)

(viii) How many elements send the line $y = 0$ to the line $x = 0$?

## EXERCISE 10

Divide a square into $n^2$ equally sized little squares as was done for $n = 3$ in Fig 1.6. Hidden in this are some $2 \times 2$ squares, e.g. the one formed by 1,2,4 and 5 in Fig 1.6, and each element of $D_4$ sends each such $2 \times 2$ square to another. In other words $D_4$ acts on the $2 \times 2$ squares. When $n = 3$ there are four $2 \times 2$ squares which form a single orbit but for larger $n$ the situation is more complex. Discuss the orbits and stabilizers of this action for different $n$. Some questions you might ask are:

(i) How many orbits are there?

(ii) What sizes can the orbits be?

(iii) What sizes can the stabilizers be? How are they related to the sizes of orbits?

Discuss what happens with $2 \times 2$ squares replaced by different shaped combinations of little squares. For example $3 \times 1$ and $1 \times 3$ rectangles (taken together because a rotation can turn a $3 \times 1$ rectangle into a $1 \times 3$ rectangle) give interesting answers when $n = 4$ and $n = 5$.

# 2 • Permutations

Many important groups, including those discussed in Chapter 1 consist of functions. The purpose of this chapter is to look at functions with an emphasis on the group properties, in particular closure and inverses. This will provide some basic examples of groups, notably the so-called symmetric groups.

## 2.1 The Symmetric Group $S_4$

In Chapter 1 we saw how certain permutations of the set $\{1, 2, 3, 4\}$ arose from symmetries of a square with numbered vertices. In this section we shall consider *all* permutations of this set. There are 24 permutations of $\{1, 2, 3, 4\}$, 24 being the number of ways of writing down the numbers 1,2,3 and 4 with no repetitions. How do we combine two permutations? We treat them as functions and combine them by composition; that is simply apply one permutation after another and see what the overall effect is. For instance, suppose

$$f = \begin{pmatrix} 1 & 2 & 3 & 4 \\ 2 & 3 & 4 & 1 \end{pmatrix} \text{ and } g = \begin{pmatrix} 1 & 2 & 3 & 4 \\ 1 & 3 & 4 & 2 \end{pmatrix}.$$

Then

$$fg(1) = f(g(1)) = f(1) = 2, \quad fg(2) = f(g(2)) = f(3) = 4,$$
$$fg(3) = f(g(3)) = f(4) = 1, \quad fg(4) = f(g(4)) = f(2) = 3.$$

Hence

$$fg = \begin{pmatrix} 1 & 2 & 3 & 4 \\ 2 & 4 & 1 & 3 \end{pmatrix}.$$

**EXERCISE 1**

Compute $gf$ with $f, g$ as above. Is $fg = gf$?

These calculations illustrate the fact that the composition of two permutations is again a permutation. As we have only considered particular choices of permutations, we have not proved this fact. It will be proved in a more general context later in the chapter.
The **identity** permutation, denoted id or $e$, is the one that leaves everything unchanged. Thus

$$\text{id} = \begin{pmatrix} 1 & 2 & 3 & 4 \\ 1 & 2 & 3 & 4 \end{pmatrix}.$$

To find the inverse of a permutation, interchange the rows. Thus, for example

$$\text{if } f = \begin{pmatrix} 1 & 2 & 3 & 4 \\ 2 & 3 & 4 & 1 \end{pmatrix} \text{ then } f^{-1} = \begin{pmatrix} 2 & 3 & 4 & 1 \\ 1 & 2 & 3 & 4 \end{pmatrix} = \begin{pmatrix} 1 & 2 & 3 & 4 \\ 4 & 1 & 2 & 3 \end{pmatrix}.$$

We have rearranged the columns so that the top row is in the usual order.

## EXERCISE 2

Find the inverses of $\begin{pmatrix} 1 & 2 & 3 & 4 \\ 4 & 3 & 1 & 2 \end{pmatrix}$ and $\begin{pmatrix} 1 & 2 & 3 & 4 \\ 3 & 4 & 1 & 2 \end{pmatrix}$.

The set of all 24 permutations of $\{1, 2, 3, 4\}$ is another example of a group and is called **symmetric group** $S_4$. The subscript 4 tells us how many numbers are being permuted.

The idea of symmetry occurs throughout mathematics, not only in the geometric context of Chapter 1. Here we look at symmetry in an algebraic context. Consider the set of all polynomials in four variables $x_1, x_2, x_3, x_4$ with integer coefficients. Three examples of such polynomials are

$$p_1 = x_1^2 + x_2^2 + x_3^2 + x_4^2, \quad p_2 = x_1x_2 + x_3 + x_4, \quad p_3 = x_1x_2^2 - x_3^3x_4^4 + 5x_1x_2x_3x_4.$$

Of these, $p_1$ has more symmetry than $p_2$ which, in turn, has more symmetry than $p_3$. What do we mean by 'symmetry' here? If we take a polynomial and permute the variables we obtain another polynomial. For example, the permutation $\begin{pmatrix} 1 & 2 & 3 & 4 \\ 2 & 3 & 4 & 1 \end{pmatrix}$ changes $p_2$ to the different polynomial $x_2x_3 + x_4 + x_1$. On the other hand, the permutation $\begin{pmatrix} 1 & 2 & 3 & 4 \\ 2 & 1 & 3 & 4 \end{pmatrix}$ sends $p_2$ to itself. The polynomial $p_1$ is unchanged by any permutation of the variables. Such a polynomial is called **symmetric**. Symmetric polynomials are important in the theory of equations and it is because $S_4$ leaves symmetric polynomials unchanged that it is called the symmetric group. In contrast, the only permutation which sends $p_3$ to itself is the identity permutation.

In Chapter 1 we looked at orbits and stabilizers of a group acting in a geometric context. Below, we investigate these ideas in the algebraic context of polynomials in four variables. The **stabilizer** of the above polynomial $p_2$ consists of those permutations which leave $p_2$ unchanged. There are four such permutations, namely

$$\text{id} = \begin{pmatrix} 1 & 2 & 3 & 4 \\ 1 & 2 & 3 & 4 \end{pmatrix}, \quad f = \begin{pmatrix} 1 & 2 & 3 & 4 \\ 1 & 2 & 4 & 3 \end{pmatrix},$$

$$g = \begin{pmatrix} 1 & 2 & 3 & 4 \\ 2 & 1 & 3 & 4 \end{pmatrix}, \quad h = \begin{pmatrix} 1 & 2 & 3 & 4 \\ 2 & 1 & 4 & 3 \end{pmatrix}.$$

The stabilizer of $p_1$ consists of all the permutations in $S_4$ and the stabilizer of $p_3$ contains only the identity permutation. The stabilizer of a polynomial is a measure of its symmetry.

The **orbit** of $p_2$ consists of all the possible polynomials reached by applying members of $S_4$ to $p_2$. There will be at most 24 such polynomials since there are 24 permutations. However, in this case there is a lot of repetition and we only obtain six *different* polynomials from $p_2$ namely

$$\begin{array}{lll} x_1x_2 + x_3 + x_4, & x_1x_3 + x_2 + x_4, & x_1x_4 + x_2 + x_3, \\ x_2x_3 + x_1 + x_4, & x_2x_4 + x_1 + x_3, & x_3x_4 + x_1 + x_2. \end{array}$$

These form the orbit of $p_2$. The orbit of the symmetric polynomial $p_1$ has only one element, $p_1$ itself, whereas the orbit of $p_3$ has 24 elements. Like the stabilizer, the orbit of a polynomial measures its symmetry but in the opposite way: the smaller the orbit, the greater the symmetry.

The next exercise investigates orbits and stabilizers for some particular polynomials. You might like to experiment further with other polynomials. Can you find a polynomial in $x_1, x_2, x_3, x_4$ with three elements in its orbit? What about five? How many different sizes of orbit can you find? Can you suggest a relationship between the size of the orbit of a polynomial and the size of its stabilizer?

**EXERCISE 3**

(i) For each of the following polynomials, how many polynomials are in its orbit and how many permutations are in its stabilizer?

(a) $x_1x_2x_3 + x_4$;

(b) $x_1x_2 + x_1x_3 + x_1x_4 + x_2x_3 + x_2x_4 + x_3x_4$;

(c) $x_1^2x_2 + x_2^2x_3 + x_3^2x_4 + x_4^2x_1$.

(ii) Draw up a Cayley table for the stabilizer $H = \{\text{id}, f, g, h\}$ of the polynomial $p_2$ in the text. Observe that $H$ is closed and that each element is its own inverse.

## 2.2 Functions

All the examples of groups which have been considered so far consist of functions. In this section we discuss functions in general.

Let $A$ and $B$ be non-empty sets. A **function** $f$ from $A$ to $B$ determines, for each $a \in A$, a *unique* element $f(a)$ of $B$. The words **mapping** and **map** are alternatives for 'function'. We use the notation $f : A \to B$ to stand for a function $f$ from $A$ to $B$. The set $A$ is called the **domain** of $f$ and $B$ is the **codomain** of $f$. Whenever we speak of a function $f : A \to B$, we shall implicitly assume that $A$ and $B$ are non-empty.

The most familiar examples of functions are probably those arising in calculus where $A$ and $B$ are subsets of the set $\mathbb{R}$ of all real numbers. The rotations $\text{rot}_\phi$ and the reflections $\text{ref}_\phi$ introduced in Chapter 1 are also functions. For these, the domain and codomain both consist of all points $P = (x, y)$ where $x, y \in \mathbb{R}$. The set of all such points is written $\mathbb{R}^2$ and is called the **Euclidean plane**. More functions from $\mathbb{R}^2$ to itself will appear in Chapter 3. The permutations discussed in Section 2.1 are functions from the set $\{1, 2, 3, 4\}$ to itself.

Two functions $f : A \to B$ and $g : C \to D$ are **equal** precisely when $A = C$, $B = D$ and $f(a) = g(a)$ for all $a \in A (= C)$.

### Composition and closure

Our discussion of the groups of symmetries in Chapter 1 and permutations in Section 2.1 emphasized not just their elements, but also the way in which they are combined using composition.

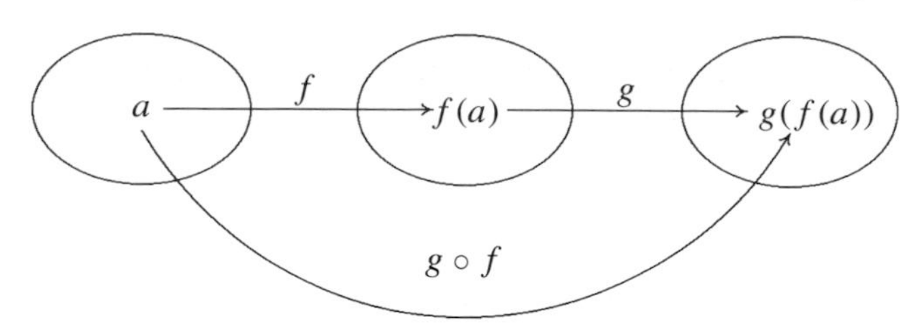


**Fig 2.1** Composition.

Let $A, B, C$ be sets and let $f : A \to B$, $g : B \to C$ be functions. The **composite** $g \circ f$ is the function $g \circ f : A \to C$ such that, for all $a \in A$, $g \circ f(a) = g(f(a))$. For example, if $A = B = C = \mathbb{R}$, $f(x) = x + 1$ and $g(x) = 2x$ for all $x \in \mathbb{R}$ then $g \circ f(x) = 2(x + 1)$ for all $x \in \mathbb{R}$. For these functions, the composite $f \circ g$ also exists, $f \circ g(x) = 2x + 1$ for all $x \in \mathbb{R}$. Observe that $g \circ f(1) \neq f \circ g(1)$ so the two composites $g \circ f$ and $f \circ g$ are different functions. It is therefore important to remember that, in working out $g \circ f$, we first apply $f$ then $g$.

Sometimes we omit the symbol $\circ$ and write $gf$ rather than $g \circ f$. This was done in Chapter 1, where we wrote $\text{rot}_\alpha \, \text{rot}_\beta = \text{rot}_{\alpha+\beta}$ rather than $\text{rot}_\alpha \circ \text{rot}_\beta = \text{rot}_{\alpha+\beta}$.

For any set $A$, the set of all functions from $A$ to $A$ is closed under composition; that is, if $f$ and $g$ are functions from $A$ to $A$ then so too is $g \circ f$.

## Associativity

Let $A, B, C, D$ be sets and let $f : A \to B$, $g : B \to C$, and $h : C \to D$ be functions. Then $g \circ f : A \to C$ so we can form the composite $h \circ (g \circ f) : A \to D$. Also, $h \circ g : B \to D$ so we can form $(h \circ g) \circ f : A \to D$. Thus, we have two functions $h \circ (g \circ f)$ and $(h \circ g) \circ f$ from $A$ to $D$. To see that they are equal, let $a \in A$. Then

$$\begin{aligned}
(h \circ (g \circ f))(a) &= h((g \circ f)(a)) \\
&= h(g(f(a))) \\
&= (h \circ g)(f(a)) \\
&= ((h \circ g) \circ f)(a).
\end{aligned}$$

Thus, we have the following **associative** law for composition of functions:

$$h \circ (g \circ f) = (h \circ g) \circ f.$$

In particular, this holds whenever $f$, $g$ and $h$ are all functions from $A$ to itself. Thus, it holds for the rotations and reflections in Chapter 1 and for the permutations in Section 2.1.

## Identity functions

Let $A$ be a set. The function from $A$ to $A$ which sends each element $a$ of $A$ to itself is called the **identity** function on $A$ and is written $\text{id}_A$. Thus, $\text{id}_A : A \to A$ and

$$\text{id}_A(a) = a \text{ for all } a \in A.$$

Now let $f : A \to B$. Then there are composites $f \circ \text{id}_A : A \to B$ and $\text{id}_B \circ f : A \to B$. But, for all $a \in A$,

$$\begin{aligned}
f \circ \text{id}_A(a) &= f(\text{id}_A(a)) = f(a) \text{ and} \\
\text{id}_B \circ f(a) &= \text{id}_B(f(a)) = f(a).
\end{aligned}$$

Thus we have

$$f \circ \mathrm{id}_A = f = \mathrm{id}_B \circ f.$$

In particular, this is true when $B = A$ and $f : A \to A$, in which case

$$f \circ \mathrm{id}_A = f = \mathrm{id}_A \circ f.$$

We say that $\mathrm{id}_A$ is **neutral** for composition.

## Inverses

Let $A, B$ be sets and let $f : A \to B$ be a function. An **inverse** for $f$ is a function $g : B \to A$ such that

$$g \circ f = \mathrm{id}_A \text{ and } f \circ g = \mathrm{id}_B .$$

That is, for all $a \in A$ and all $b \in B$,

$$g(f(a)) = a \text{ and } f(g(b)) = b.$$

A function is said to be **invertible** if it has an inverse.

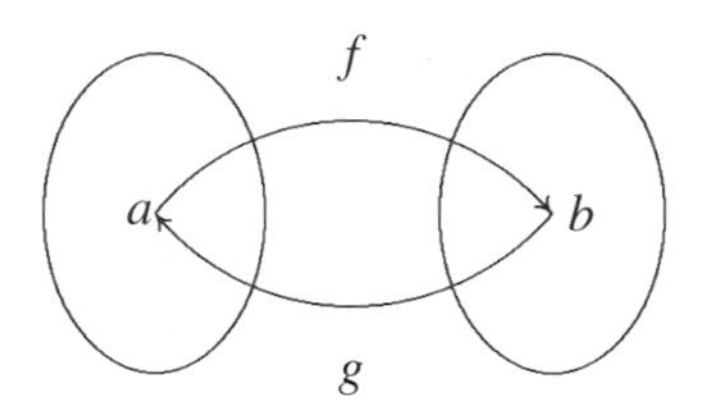

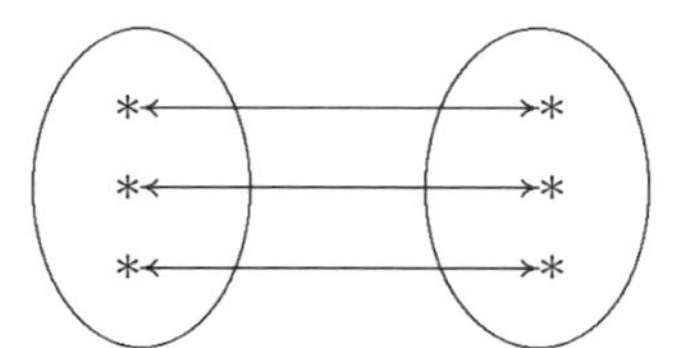

**Fig 2.2** Inverse functions.

For an easy example, let $f, g : \mathbb{R} \to \mathbb{R}$ be the functions such that $f(x) = x + 2$ and $g(x) = x - 2$ for all $x \in \mathbb{R}$. Then $g$ and $f$ are inverses of each other. Also, for any set $A$, the identity function $\mathrm{id}_A$ is an inverse for itself. Consider now the functions $\mathrm{rot}_\theta : \mathbb{R}^2 \to \mathbb{R}^2$ and $\mathrm{ref}_\theta : \mathbb{R}^2 \to \mathbb{R}^2$. Note that $\mathrm{rot}_0$ is simply the identity function $\mathrm{id}_{\mathbb{R}^2}$. The formulae in Table 1.1 then show that $\mathrm{rot}_\theta$ has inverse $\mathrm{rot}_{-\theta} (= \mathrm{rot}_{2\pi-\theta})$ and that $\mathrm{ref}_\theta$ is its own inverse. Hence, these rotations and reflections are invertible functions from $\mathbb{R}^2$ to itself.

There is no guarantee that a given function $f : A \to B$ has an inverse $g$. For example, $f : \mathbb{R} \to \mathbb{R}$ where $f(x) = x^2$ does not have an inverse. There are two reasons for this. Suppose that $f$ has inverse $g : \mathbb{R} \to \mathbb{R}$. Firstly, $-1 \in \mathbb{R}$ so we would have an element $x = g(-1) \in \mathbb{R}$. But then $x^2 = f(x) = f(g(-1)) = f \circ g(-1) = \mathrm{id}(-1) = -1$, which is impossible with $x \in \mathbb{R}$. Secondly, $f(-1) = f(1)$ so, applying $g$ we get $g \circ f(-1) = g \circ f(1)$. Since $g \circ f = \mathrm{id}$, this says $-1 = 1$ which is absurd. Thus, there are two obstructions to the existence of an inverse:

- there is no element $x \in \mathbb{R}$ with $f(x) = -1$;
- there are two different elements $a_1 = 1$ and $a_2 = -1$ with $f(a_1) = f(a_2)$.

This leads us to make the following definitions, where $A$, $B$ are sets and $f : A \to B$ is a function.

To say that $f$ is **surjective**, or a **surjection**, means that, for each $b \in B$, there exists at least one element $a \in A$ such that $f(a) = b$.

To say that $f$ is **injective**, or an **injection**, means that, for each $b \in B$, there exists at most one element $a \in A$ such that $f(a) = b$. This is equivalent to saying that if $a_1, a_2 \in A$ are such that $f(a_1) = f(a_2)$ then $a_1 = a_2$.

A function which is both injective and surjective is said to be **bijective**, or a **bijection**. Thus, to say that $f$ is bijective means that, for each $b \in B$, there is a unique element $a \in A$ such that $f(a) = b$.

Thus, the function $f : \mathbb{R} \to \mathbb{R}$ with $f(x) = x^2$ is neither injective nor surjective. However, if $B$ denotes the set of all positive real numbers then the function $h : B \to B$ with $h(x) = x^2$ is bijective and has inverse $g : B \to B$ given by $g(y) = \sqrt{y}$ for all $y \in B$.

## • Theorem 1

A function $f : A \to B$ has an inverse if and only if $f$ is bijective. ●

PROOF

We need to prove two things, an 'if' part and an 'only if' part. For the 'if' part, we must show that if $f$ is bijective then $f$ has an inverse $g : B \to A$. Suppose that $f$ is bijective. For each $b \in B$, let $g(b)$ be the *unique* element $a \in A$ with $f(a) = b$. Such an element exists since $f$ is bijective. This defines a function $g : B \to A$ and it is clear that, for all $b \in B$, $f(g(b)) = b$, that is $f \circ g = \mathrm{id}_B$. Setting $b = f(a)$ in the definition of $g$, we see that $g(f(a)) = a$ for all $a \in A$ and so $g \circ f = \mathrm{id}_A$. Thus, $g$ is an inverse of $f$.

For the 'only if' part, we must show that $f : A \to B$ has an inverse $g : B \to A$ only if it is bijective. In other words, $f$ cannot have an inverse without being bijective. This is the same as saying that if $f$ has an inverse then it is bijective. So suppose that $f$ has an inverse $g : B \to A$. We shall show that $f$ must be bijective. For surjectivity, let $b \in B$ and set $a = g(b) \in A$. Then $f(a) = f(g(b)) = f \circ g(b) = \mathrm{id}_B(b) = b$. Thus, $f$ is surjective. For injectivity, let $a_1, a_2 \in A$ be such that $f(a_1) = f(a_2)$. We must show that $a_1 = a_2$. Applying $g$ to both sides of the equation $f(a_1) = f(a_2)$, we obtain $g(f(a_1)) = g(fa_2))$. But $g \circ f = \mathrm{id}_A$ so $g(f(a_1)) = a_1$, $g(f(a_2)) = a_2$ and hence $a_1 = a_2$. Therefore, $f$ is injective and, as we have already shown it to be surjective, it is bijective. ●

Let $f : A \to B$ be an invertible function. Then $f$ cannot have two different inverses because, if $g : B \to A$ and $h : B \to A$ are inverses of $f$ and $b \in B$, then $b = f(g(b)) = f(h(b))$ and since $f$ is injective $g(b) = h(b)$. The unique inverse of $f : A \to B$ is written $f^{-1} : B \to A$. Since $f \circ f^{-1} = \mathrm{id}_B$ and $f^{-1} \circ f = \mathrm{id}_A$, $f$ is the inverse of $f^{-1}$ and so $f^{-1}$ is also invertible.

A bijective function $f : A \to B$, together with its inverse, pairs off the elements of $A$ with those of $B$, see Fig 2.2. In particular, if $A = \{a_1, a_2, \ldots, a_n\}$ has only a finite number, $n$, of different elements and there is a bijection $f : A \to B$ then, by surjectivity, $f(a_1), f(a_2), \ldots, f(a_n)$ are all the elements of $B$ and, by injectivity, they are all different. So $B$ also has exactly $n$ elements.

## Closure of invertible functions

Let $X$ be a set. We have noted that the set of all functions from $X$ to itself is closed under composition. In order to form a group of functions, we require an inverse for each function and so we need to consider only invertible functions from $X$ to $X$. We need to check the closure of the set of invertible functions under composition. The following exercise illustrates the idea.

## EXERCISE 4

Let $f : \mathbb{R} \to \mathbb{R}$ and $g : \mathbb{R} \to \mathbb{R}$ be the functions such that $f(x) = x - 3$ and $g(x) = 2x$.

(i) Find inverses $f^{-1}$ and $g^{-1}$ for $f$ and $g$.

(ii) Compute the composites $fg$, $gf$, $f^{-1}g^{-1}$ and $g^{-1}f^{-1}$.

(iii) Compute $(fg)(f^{-1}g^{-1})$, $(fg)(g^{-1}f^{-1})$, $(f^{-1}g^{-1})(fg)$ and $(g^{-1}f^{-1})(fg)$. Suggest a rule for the inverse of $fg$. What is the inverse of $gf$?

### ● *Theorem 2*

Let $f : X \to X$ and $g : X \to X$ be invertible functions. Then $fg$ is invertible and $(fg)^{-1} = g^{-1}f^{-1}$. ●

PROOF

Consider the composite $g^{-1}f^{-1} : X \to X$. We shall show that this is an inverse for $fg$. Using associativity of composition and neutrality of $\mathrm{id}_X$

$$\begin{aligned}(fg)(g^{-1}f^{-1}) &= f(gg^{-1})f^{-1} = ff^{-1} = \mathrm{id}_X \text{ and} \\ (g^{-1}f^{-1})(fg) &= g^{-1}(f^{-1}f)g = g^{-1}g = \mathrm{id}_X .\end{aligned}$$

Thus, $fg$ has inverse $g^{-1}f^{-1}$. ●

## EXERCISE 5

Complete the following table showing whether the given functions are surjective/injective/bijective. For a non-surjective function enter an element $b \in B$ such that $f(a) \neq b$ for all $a \in A$ in the appropriate column. For a non-injective function, enter two different elements $a_1, a_2$ of $A$ with $f(a_1) = f(a_2)$. In the last two lines $O_2$ stands for the orthogonal group introduced in Chapter 1.

| Domain <br> $A$ | Codomain <br> $B$ | $f(a)$ | surj? <br> yes/no | inj? <br> yes/no | bij? <br> yes/no | $b$ with, for all $a$, $f(a) \neq b$ | $a_1, a_2$ with $a_1 \neq a_2$ $f(a_1) = f(a_2)$ |
|---|---|---|---|---|---|---|---|
| $\mathbb{R}$ | $\mathbb{R}$ | $a^2$ | no | no | no | $-1$ | $-1, 1$ |
| $\{x \in \mathbb{R} : x \geqslant 0\}$ | $\{x \in \mathbb{R} : x \geqslant 0\}$ | $a^2$ | yes | yes | yes | none | none |
| $\{x \in \mathbb{R} : x \geqslant 0\}$ | $\mathbb{R}$ | $a^2$ | | | | | |
| $\mathbb{R}$ | $\{x \in \mathbb{R} : x \geqslant 0\}$ | $a^2$ | | | | | |
| $\mathbb{R}$ | $\mathbb{R}$ | $a^3$ | | | | | |
| $\mathbb{R}$ | $\mathbb{R}$ | $a^4$ | | | | | |
| $\mathbb{R}\backslash\{0\}$ | $\mathbb{R}\backslash\{0\}$ | $\frac{1}{a}$ | | | | | |
| $\mathbb{R}$ | $O_2$ | $\text{rot}_a$ | | | | | |
| $\mathbb{R}$ | $O_2$ | $\text{ref}_a$ | | | | | |

## 2.3 Permutations

Whereas a bijective function between two different sets pairs off the elements of the two sets, a bijective function from a set to itself rearranges or permutes the elements of the set. For this reason a bijective function (or, equivalently, an invertible function) from a set $X$ to itself is called a **permutation** of $X$. The set of all permutations of $X$ will be denoted by $S_X$. If $X = \{1, 2, \ldots, n\}$ is the set of the first $n$ positive integers then we shall write $S_n$ rather than $S_X$. The theory of permutations of an arbitrary finite set $X = \{x_1, x_2, \ldots, x_n\}$ can be obtained from that of $\{1, 2, \ldots, n\}$ by replacing 1 by $x_1$ etc.

### Group properties of $S_X$ and $S_n$

Let $X$ be a set. The set $S_X$ of all permutations satisfies the four group properties:

- closure: $fg \in S_X$ for all $f, g \in S_X$;
- associativity: $(fg)h = f(gh)$ for all $f, g, h \in S_X$ ;
- existence of a neutral element: the identity function $\text{id}_X \in S_X$ and is neutral for composition;
- existence of inverses: for each $f \in S_X$ there exists $g \in S_X$ such that $fg = \text{id}_X = gf$.

In particular, these properties hold when $S_X = S_n$ is the set of all permutations of $\{1, 2, \ldots, n\}$.

### Permutation notation

To specify a permutation $f$ of $\{1, 2, \ldots, n\}$ it is usually pointless to give a formula for $f(i)$ in general. Rather, one needs to specify $f(i)$ for each value of $i$, $1 \leqslant i \leqslant n$. For this reason, the two-row notation, which we have already used in Chapter 1 and Section 2.1, is commonly used. In this notation an arbitrary permutation $f$ is written

$$f = \begin{pmatrix} 1 & 2 & \ldots & n \\ f(1) & f(2) & \ldots & f(n) \end{pmatrix}.$$

In particular

$$\text{id} = \begin{pmatrix} 1 & 2 & \ldots & n \\ 1 & 2 & \ldots & n \end{pmatrix}$$

and the permutation which reverses the order of $1, 2, \ldots, n$ is

$$\begin{pmatrix} 1 & 2 & \ldots & n \\ n & n-1 & \ldots & 1 \end{pmatrix}.$$

Each of the numbers from 1 to $n$ must appear exactly once in the bottom row of a permutation $f$ in $S_n$. There are $n$ possibilities for $f(1)$ and for each of these there are $n-1$ possibilities for $f(2)$. There are then $n-2$ possibilities for $f(3)$, $n-3$ for $f(4), \ldots, 1$ for $f(n)$. Thus, $S_n$ has $n!$ (that is $n(n-1)(n-2)\ldots 1$) elements.

In Section 2.1 we have calculated some composites of permutations and some inverses. The first of the exercises below is intended to revise such calculations.

**EXERCISE 6**

Let $f = \begin{pmatrix} 1 & 2 & 3 & 4 & 5 \\ 5 & 3 & 1 & 4 & 2 \end{pmatrix}$ and $g = \begin{pmatrix} 1 & 2 & 3 & 4 & 5 \\ 4 & 1 & 5 & 3 & 2 \end{pmatrix}$. Calculate $fg, gf, f^{-1}$ and $f^{-1}g$. (Remember, when calculating $fg$ to apply $g$ first then $f$.)

**EXERCISE 7**

List all six elements of $S_3$ in two-row notation.

**EXERCISE 8**

How many elements $f$ of $S_5$ satisfy the equation $f(5) = 5$?

## Cycles

The two-row notation for permutations does not easily give much insight into the nature of a particular permutation. Here, we introduce an alternative one-row notation for certain permutations called **cycles**. We will later see how to write any permutation in terms of cycles. This is the so-called **cycle decomposition** and for some purposes it is preferable to the two-row notation. Consider the permutation

$$f = \begin{pmatrix} 1 & 2 & 3 & 4 \\ 2 & 3 & 4 & 1 \end{pmatrix} \in S_4.$$

Then $f$ sends 1 to 2, 2 to 3, 3 to 4 and 4 to 1. If the numbers 1 to 4 are arranged anticlockwise in a circle then $f$ moves each number one place anticlockwise round the circle. This is an example of a **cyclic** permutation or **cycle**. It is written $(1\ 2\ 3\ 4)$. Here, each number, except the last, is moved one place to the right and the last is moved to the front.

Now let $n$ and $k$ be positive integers with $k \leqslant n$ and let $a_1, a_2, \ldots a_k$ be $k$ *distinct* elements of $\{1, 2, \ldots, n\}$. The permutation $f \in S_n$ such that

$$f(a_1) = a_2, \; f(a_2) = a_3, \ldots, f(a_{k-1}) = a_k, \; f(a_k) = a_1,$$

and $f(a) = a$ for each $a$ not appearing in the list $a_1, a_2, \ldots, a_k$, is called a **cycle** of

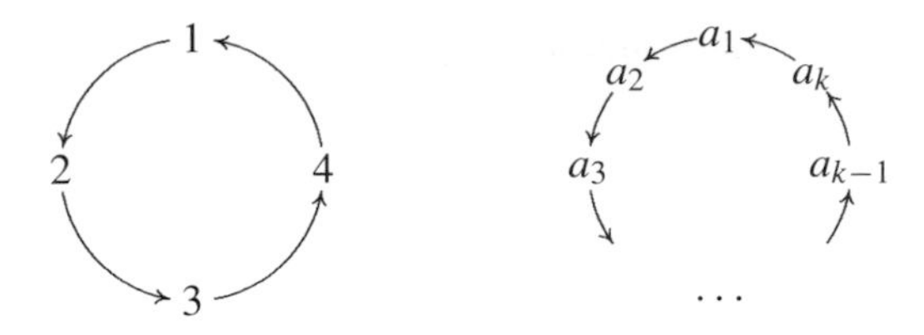


**Fig 2.3** Cycles.

**length** $k$ or a $k$-**cycle**. It is written

$$(a_1\ a_2\ \dots\ a_{k-1}\ a_k).$$

For example, in $S_7$,

$$(1\ 3\ 5\ 7\ 2) = \begin{pmatrix} 1 & 2 & 3 & 4 & 5 & 6 & 7 \\ 3 & 1 & 5 & 4 & 7 & 6 & 2 \end{pmatrix}.$$

Notice that as 4 and 6 do not appear in the cycle, they are sent to themselves. The existence of two different notations for permutations can be confusing and we strongly advise you to spend time on the exercise below. This illustrates general ideas that we will discuss in the next section.

## EXERCISE 9

(i) Write each of the following cycles in two-row notation:

(a) $(1\ 2\ 3)$ and $(2\ 3\ 1)$ in $S_3$;

(b) $(1\ 2\ 3\ 4\ 5)$ and $(5\ 4\ 3\ 2\ 1)$ in $S_5$.

(ii) In $S_3$, let $f = (1\ 2\ 3)$. Compute $f^2$ and $f^3$. Suggest a rule for $f^k$ when $f$ is a $k$-cycle.

(iii) In $S_5$, let $f = (1\ 2\ 3\ 4\ 5)$ and $g = (5\ 4\ 3\ 2\ 1)$. Compute $fg$ and $gf$. Suggest a rule for the inverse of a cycle.

(iv) In $S_6$, let $f = (1\ 2\ 4)$ and $g = (3\ 5\ 6)$. Compute $fg$. Is $gf = fg$?

(v) In $S_6$, let $h = (1\ 2\ 3)$ and $g = (3\ 5\ 6)$. Compute $hg$. Is $gh = hg$?

## 2.4 Basic Properties of Cycles

In this section we list some basic properties of cycles which you may have observed in Exercise 9.

In Exercise 9(i), $(1\ 2\ 3) = (2\ 3\ 1)$. This 3-cycle is also equal to $(3\ 1\ 2)$.

- A $k$-cycle can be written in $k$ different ways, each with the same 'cyclic order'. For example $(1\ 7\ 3\ 4) = (7\ 3\ 4\ 1) = (3\ 4\ 1\ 7) = (4\ 1\ 7\ 3)$.

In Exercise 9(ii), if $f = (1\ 2\ 3)$ then $f^2 = (1\ 3\ 2)$ and $f^3 = \text{id}$. In general, if $f = (a_1\ a_2 \dots a_k)$ is a cycle of length $k$ then $f^k = \text{id}$, because $f^k$ moves each $a_i$ around the circle $k$ places, that is back to itself. Also, if $1 \leqslant j < k$ then $f^j \neq \text{id}$.

- If $f$ is a $k$-cycle then $k$ is the *least* positive integer such that $f^k = \mathrm{id}$.

The effect of a cycle $f$ is to move each $a_i$ one place anticlockwise round the circle. The effect of $f^{-1}$ must be to move each $a_i$ one place clockwise round the circle. This is the same as moving each $a_i$ anticlockwise $k - 1$ places.

- If $f = (a_1\ a_2 \ldots a_k)$ is a $k$-cycle then $f^{-1} = (a_k \ldots a_2\ a_1) = f^{k-1}$.

A cycle $(a_1)$ of length 1 sends $a_1$ to itself and also sends each other number to itself.

- A cycle of length 1 is the identity.

Two cycles $(a_1\ a_2 \ldots a_k)$ and $(b_1\ b_2 \ldots b_l)$ are said to be **disjoint** if $a_i \neq b_j$ for all $i$ and $j$. For example $f = (1\ 2\ 4)$ and $g = (3\ 5\ 6)$ are disjoint but $h = (1\ 2\ 3)$ and $g$ are not. The two disjoint cycles $f, g$ **commute** but $g$ and $h$ do not commute. That is, $fg = gf$ but $hg \neq gh$.

Let $f = (a_1\ a_2 \ldots a_k)$ and $g = (b_1\ b_2 \ldots b_l)$ be disjoint cycles in $S_n$. We now show that $fg = gf$. To do so, we have to show that $fg(a) = gf(a)$ for each $a$, $1 \leqslant a \leqslant n$. There are three cases: $a$ can appear in $f$, or in $g$, or in neither. If $a$ appears in $f$ then we can rearrange the numbers appearing in $f$ cyclically so that $a = a_1$. As $a_1$ and $a_2$ do not appear in $g$, we have $g(a_1) = a_1$ and $g(a_2) = a_2$. Hence, $gf(a_1) = g(a_2) = a_2$ and $fg(a_1) = f(a_1) = a_2$ and so $fg(a) = a_2 = gf(a)$. Similarly, if $a$ appears in $g$, we can assume that $a = b_1$. Then $fg(a) = b_2 = gf(a)$. Finally, if $a$ appears in neither $f$ nor $g$ then $f(a) = a = g(a)$ and so $fg(a) = a = gf(a)$.

- Disjoint cycles commute.

## 2.5 Cycle Decomposition

In this section we shall see how to write an arbitrary permutation $f$ as a product of disjoint cycles. This gives the so-called **cycle decomposition** of $f$. For example

$$(1\ 7\ 9\ 3)(2\ 4\ 8\ 6)$$

is the cycle decomposition of the permutation

$$\begin{pmatrix} 1 & 2 & 3 & 4 & 5 & 6 & 7 & 8 & 9 \\ 7 & 4 & 1 & 8 & 5 & 2 & 9 & 6 & 3 \end{pmatrix}$$

of the nine squares in Fig 1.6 given by the rotation $\mathrm{rot}_{\frac{\pi}{2}}$. Notice that the cycle decomposition gives a much better feel for the effect of this permutation, namely that it simultaneously cyclically permutes the four 'corners' 1, 7, 3, 9 and the four 'midpoints' 2, 4, 8, 6.

The following algorithm will express any permutation $f \in S_n$ as a product of disjoint cycles. We illustrate its use with the permutation

$$f = \begin{pmatrix} 1 & 2 & 3 & 4 & 5 & 6 & 7 & 8 & 9 & 10 & 11 & 12 \\ 8 & 6 & 1 & 9 & 10 & 2 & 7 & 3 & 12 & 5 & 4 & 11 \end{pmatrix}.$$

## Step 1

Write out $1, f(1), f^2(1), f^3(1), \ldots$ until $f^{k_1}(1) = 1$ for some positive integer $k_1$. Let $f_1$ be the cycle $(1\ f(1)\ f^2(1)\ \ldots\ f^{k_1-1}(1))$. In the example $f_1 = (1\ 8\ 3)$. Then $f$ and $f_1$ have the same effect on the numbers appearing in $f_1$. If $1, 2, \ldots, n$ all appear in $f_1$ then $f = f_1$ and the algorithm stops.

## Step 2

Choose the smallest integer $j$, $1 \leqslant j \leqslant n$, *not* appearing in $f_1$. In the example $j = 2$. Write out $j, f(j), f^2(j), f^3(j), \ldots$ until $f^{k_2}(j) = j$ for some positive integer $k_2$. Let $f_2$ be the cycle $(j\ f(j)\ f^2(j)\ \ldots\ f^{k_2-1}(j))$. In the example $f_2 = (2\ 6)$. No number appearing in $f_2$ can appear in $f_1$ otherwise all numbers appearing in $f_2$, including $j$, would appear in $f_1$. The permutations $f$ and $f_1 f_2$ have the same effect on the numbers appearing in $f_1$ or $f_2$. If $1, 2, \ldots, n$ all appear in $f_1$ or $f_2$ then $f = f_1 f_2$ and the algorithm stops.

## Step 3

Repeat Step 2 with $j$ now the smallest integer not appearing in $f_1$ or $f_2$ to find $f_3$. In the example $f_3 = (4\ 9\ 12\ 11)$ The algorithm stops after, say, $s$ steps with $f = f_1 f_2 \ldots f_s$. This is the **cyclic decomposition** of $f$. In the example, $s = 5$ and $f = (1\ 8\ 3)(2\ 6)(4\ 9\ 12\ 11)(5\ 10)(7)$.

To what extent is the cycle decomposition of a permutation unique? If we had given priority to larger numbers in our algorithm above we would have obtained

$$f = (12\ 11\ 4\ 9)(10\ 5)(8\ 3\ 1)(7)(6\ 2).$$

This looks different but is equally valid. The differences are both explained by observations we have already made. Firstly, disjoint cycles commute so, in any product of disjoint cycles the order of the cycles can be changed without affecting the product. Secondly, each cycle can be written in different ways, but each with the same cyclic order. There is one further change that can be made to the cyclic decomposition. Any cycle of length 1, such as (7) above, is the identity permutation in disguise, and so can be deleted from the product. With these exceptions, the expression for a permutation as a product of disjoint cycles is unique.

Sometimes we wish to take a product $p$ of non-disjoint cycles and rewrite it as a product of disjoint cycles. To do this we follow the above algorithm but there is no need to rewrite $p$ in two-row notation first. For example, let $p = (1\ 2\ 3)(4\ 5)(1\ 2)(3\ 4\ 5)$. Start with 1 as usual and remember to apply the cycles in the product from the right.

$$1 \underbrace{\mapsto}_{(3\ 4\ 5)} 1 \underbrace{\mapsto}_{(1\ 2)} 2 \underbrace{\mapsto}_{(4\ 5)} 2 \underbrace{\mapsto}_{(1\ 2\ 3)} 3.$$

Thus, $p(1) = 3$ and, to continue the algorithm, we next find $p(3)$. This is done in the same way as $p(1)$ and gives $p(3) = 5$. Similarly, $p(5) = 1$, completing a cycle. Next find $p(2)$. You should find $p(2) = 2$ and $p(4) = 4$. So $p = (1\ 3\ 5)(2)(4) = (1\ 3\ 5)$ is, in fact, a cycle.

### EXERCISE 10

(i) Find the cycle decomposition of the permutation

$$\begin{pmatrix} 1 & 2 & 3 & 4 & 5 & 6 & 7 & 8 & 9 & 10 \\ 2 & 9 & 7 & 1 & 5 & 10 & 8 & 3 & 4 & 6 \end{pmatrix}.$$

(ii) Find the cycle decomposition of each of the following products of non-disjoint cycles:

(a) (1 2 3)(2 3 4)(3 4 5);

(b) (1 2)(2 3)(3 4)(4 5);

(c) (1 5)(1 4)(1 3)(1 2);

(d) (1 2)(2 3)(3 4)(2 3)(1 2).

## 2.6 Transpositions

A cycle $(i\ j)$ of length 2 simply interchanges or transposes $i$ and $j$ and is called a **transposition**.

Suppose we wish to rearrange

♠ ♡ ◇ ♣ as ♣ ♠ ♡ ◇

using only transpositions. This corresponds to expressing the cycle (1 2 3 4) as a product of transpositions. One way is to move ♣ past each of the others in turn as shown below.

| | 1 | 2 | 3 | 4 |
|---|---|---|---|---|
| | ♠ | ♡ | ◇ | ♣ |
| apply (3 4): | ♠ | ♡ | ♣ | ◇ |
| apply (2 3): | ♠ | ♣ | ♡ | ◇ |
| apply (1 2): | ♣ | ♠ | ♡ | ◇ |

This shows that

$$(1\ 2\ 3\ 4) = (1\ 2)(2\ 3)(3\ 4).$$

Remember that we start at the right-hand end of the product. A general formula is

$$(a_1\ a_2\ a_3 \ldots a_k) = (a_1\ a_2)(a_2\ a_3) \ldots (a_{k-1}\ a_k).$$

An alternative approach is shown below.

| | 1 | 2 | 3 | 4 |
|---|---|---|---|---|
| | ♠ | ♡ | ◇ | ♣ |
| apply (1 2): | ♡ | ♠ | ◇ | ♣ |
| apply (1 3): | ◇ | ♠ | ♡ | ♣ |
| apply (1 4): | ♣ | ♠ | ♡ | ◇ |

Thus, (1 2 3 4) = (1 4)(1 3)(1 2). This illustrates an alternative general formula, namely

$$(a_1\ a_2 \ldots a_k) = (a_1\ a_k)(a_1\ a_{k-1}) \ldots (a_1\ a_2).$$

Any permutation $f \in S_n$, $n \geqslant 2$, can be expressed as a product of transpositions as follows.

If $f = \text{id}$ then $f = (1\ 2)(1\ 2)$. If $f \neq \text{id}$ then first express $f$ as a product of disjoint cycles, deleting any cycles of length one. Then apply one of the formulae above (whichever you find easier to remember) to replace each cycle by a product of transpositions. In the example we used to illustrate the cycle decomposition algorithm,

$$\begin{aligned} f &= (1\ 8\ 3)(2\ 6)(4\ 9\ 12\ 11)(5\ 10)(7) \\ &= (1\ 8)(8\ 3)(2\ 6)(4\ 9)(9\ 12)(12\ 11)(5\ 10) \\ \text{or} \quad &= (1\ 3)(1\ 8)(2\ 6)(4\ 11)(4\ 12)(4\ 9)(5\ 10) \end{aligned}$$

## Adjacent transpositions

Suppose we wish to rearrange

♠ ♡ ◇ ♣ as ♣ ♡ ◇ ♠

using only transpositions involving adjacent numbers $i$ and $i+1$. This corresponds to expressing the transposition $(1\ 4)$ as a product of the transpositions $(1\ 2)$, $(2\ 3)$, $(3\ 4)$.

| | 1 | 2 | 3 | 4 |
|---|---|---|---|---|
| | ♠ | ♡ | ◇ | ♣ |
| apply (1 2): | ♡ | ♠ | ◇ | ♣ |
| apply (2 3): | ♡ | ◇ | ♠ | ♣ |
| apply (3 4): | ♡ | ◇ | ♣ | ♠ |
| apply (2 3): | ♡ | ♣ | ◇ | ♠ |
| apply (1 2): | ♣ | ♡ | ◇ | ♠ |

Thus, $(1\ 4) = (1\ 2)(2\ 3)(3\ 4)(2\ 3)(1\ 2)$.

A general formula, which holds when $i < j$, is

$$(i\ j) = (i\ i+1)(i+1\ i+2)\ldots(j-2\ j-1)(j-1\ j)(j-2\ j-1)\ldots(i+1\ i+2)(i\ i+1).$$

Here $(i\ j)$ is expressed as the product of the odd number $2(j-i)-1$ of transpositions.

## EXERCISE 11

Express the permutation $\begin{pmatrix} 1 & 2 & 3 & 4 & 5 & 6 & 7 & 8 & 9 & 10 \\ 2 & 9 & 7 & 1 & 5 & 10 & 8 & 3 & 4 & 6 \end{pmatrix}$ as a product of transpositions.

## EXERCISE 12

(i) Find a general formula expressing $(i\ j)$ as a product of transpositions involving 1. (How would you rearrange

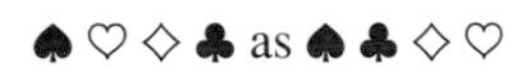

using only transpositions involving the left-hand position? This corresponds to expressing the transposition $(2\ 4)$ as a product of transpositions involving 1.)

(ii) Using the formula in (i), every permutation in $S_n$, $n \geqslant 2$, can be written as a product of transpositions of the form $(1\ c)$. Write the permutation

$$\begin{pmatrix} 1 & 2 & 3 & 4 & 5 & 6 & 7 & 8 & 9 & 10 \\ 2 & 9 & 7 & 1 & 5 & 10 & 8 & 3 & 4 & 6 \end{pmatrix}$$

as such a product.

## 2.7 The 15-puzzle

You will be aware by now that there is little uniqueness about the expression for a permutation as a product of transpositions. There were two alternative formulae for a cycle as a product of transpositions and each transposition could then be replaced by a product of adjacent transpositions or by a product of transpositions involving 1. However, there is one thing which stays fixed throughout these possible changes and that is whether the number of transpositions in the product is even or odd. The two alternatives for a $k$-cycle each involve $k - 1$ transpositions while in Exercise 2.11 your formula should involve replacing one transposition by three. In the formula for adjacent transpositions, one transposition is replaced by $2(j - i) - 1$ transpositions. This raises the question, can a product of an odd number of transpositions equal the product of an even number of transpositions? Certainly the formulae we have used will not produce such a possibility but perhaps we have not found the right formula. One special case of this question is the basis for a famous puzzle.

The 15-puzzle is a tray with 15 numbered tiles, as illustrated below, which are free to slide into the single available blank space but not to be removed from the tray. The idea is to find a sequence of moves converting the tray from its position on the left to its position on the right.

| 1 | 2 | 3 | 4 |
|---|---|---|---|
| 5 | 6 | 7 | 8 |
| 9 | 10 | 11 | 12 |
| 13 | 15 | 14 | |

| 1 | 2 | 3 | 4 |
|---|---|---|---|
| 5 | 6 | 7 | 8 |
| 9 | 10 | 11 | 12 |
| 13 | 14 | 15 | |

**Fig 2.4** 15-puzzle.

In other words, the idea is to perform the transposition (14 15) by a sequence of permissible moves, each of which is a transposition involving the latest position of the blank. For example, we might begin with (14 16) and follow with (11 14). If the squares are alternatively coloured black and white, as in a chessboard, the colour of the blank space changes on each move. Now our required overall move has the blank in the same position, and hence of the same colour, before and after. So, if it is possible it must involve an *even* number of transpositions. Thus, the problem amounts to expressing one transposition as a product of an even number of transpositions. We leave you to ponder this but shall return to it at the end of Chapter 7.

## 2.8 Further Exercises on Chapter 2

### EXERCISE 13

The set of permutations on $\{1, 2, 3\}$ form the symmetric group $S_3$. This group acts on polynomials in $x_1, x_2, x_3$. This question investigates this action.

(i) How many elements does $S_3$ have?

(ii) What are the orbit and stabilizer of the polynomial $x_1x_2 + x_3$?

(iii) What are the orbit and stabilizer of the polynomial $x_1x_2^2 + x_2x_3^2 + x_3x_1^2$?

(iv) Can you suggest a relationship between the number of elements in $S_3$, the number of elements in the orbit of a particular polynomial and the number of elements in the stabilizer of the same polynomial? Experiment with a few polynomials of your own choosing.

### EXERCISE 14

Let $f : \mathbb{R} \to \mathbb{R}$ and $g : \mathbb{R} \to \mathbb{R}$ be the functions such that, for all $a \in \mathbb{R}$, $f(a) = 3a^2$ and $g(a) = a + 2$.

(i) Find $(f \circ g)(a)$ and $(g \circ f)(a)$ for all $a \in \mathbb{R}$.

(ii) Find a function $h : \mathbb{R} \to \mathbb{R}$ such that $f \circ g = g \circ h$.

### EXERCISE 15

(i) There are $8\ (= 2^3)$ different functions from $\{1, 2, 3\}$ to $\{1, 2\}$. How many of these are surjective? How many are injective?

(ii) There are $9\ (= 3^2)$ different functions from $\{1, 2\}$ to $\{1, 2, 3\}$. How many of these are surjective? How many are injective?

### EXERCISE 16

Find the cycle decomposition of each of the following permutations:

(i) $\begin{pmatrix} 1 & 2 & 3 & 4 & 5 & 6 & 7 & 8 & 9 & 10 \\ 1 & 3 & 5 & 7 & 9 & 8 & 4 & 10 & 2 & 6 \end{pmatrix}$;

(ii) $\begin{pmatrix} 1 & 2 & 3 & 4 & 5 & 6 & 7 & 8 & 9 & 10 \\ 4 & 2 & 5 & 8 & 6 & 1 & 3 & 7 & 9 & 10 \end{pmatrix}$.

Express each of these permutations as a product of transpositions and as a product of adjacent transpositions.

## EXERCISE 17

Let $f = \begin{pmatrix} 1 & 2 & 3 & 4 & 5 & 6 \\ 2 & 4 & 5 & 6 & 3 & 1 \end{pmatrix}$ and $g = \begin{pmatrix} 1 & 2 & 3 & 4 & 5 & 6 \\ 5 & 6 & 3 & 1 & 4 & 2 \end{pmatrix}$. Find $fg$, $gf$ and $f^{-1}g$.

## EXERCISE 18

Find the cycle decomposition of each of the following products of non-disjoint cycles:

(i) $(3\ 4\ 5)(2\ 3\ 4)(1\ 2\ 3)$;

(ii) $(1\ 2\ 4\ 5)(1\ 2\ 4)(1\ 2)$.

## EXERCISE 19

Find the cycle decomposition of $f^2$ where

(i) $f = (1\ 2\ 3 \ldots m\ \ m+1 \ldots 2m-1\ \ 2m)$, a cycle of even length $2m$;

(ii) $f = (1\ 2\ 3 \ldots m\ \ m+1 \ldots 2m\ \ 2m+1)$, a cycle of odd length $2m+1$.

## EXERCISE 20

For $n \in \mathbb{N}$, let $f_n = \begin{pmatrix} 1 & 2 & \ldots & \ldots & n \\ n & n-1 & \ldots & \ldots & 1 \end{pmatrix}$. For example, $f_4 = \begin{pmatrix} 1 & 2 & 3 & 4 \\ 4 & 3 & 2 & 1 \end{pmatrix}$ and $f_5 = \begin{pmatrix} 1 & 2 & 3 & 4 & 5 \\ 5 & 4 & 3 & 2 & 1 \end{pmatrix}$. Find the cycle decomposition of $f_n$, distinguishing between the cases $n$ even or odd.

# 3 • Linear Transformations and Matrices

In Chapter 1 we discussed rotations and reflections which are functions from $\mathbb{R}^2$ to itself. These are two of a general class of functions called *linear transformations*. We shall introduce linear transformations in terms of matrices and matrix multiplication. Although matrices will be familiar to many readers, we give a brief introduction in this chapter emphasizing those points that are relevant to the groups we wish to discuss.

## 3.1 Matrix Multiplication

An $m \times n$ **matrix** over $\mathbb{R}$ is an array of real numbers with $m$ rows and $n$ columns. For example, $\begin{pmatrix} 2 & 1 & 4 \\ 3 & -1 & 0 \end{pmatrix}$ is a $2 \times 3$ matrix. For our purposes the important matrices are the **square** matrices; that is, matrices which are $n \times n$ for some $n$. The set of all $n \times n$ matrices over $\mathbb{R}$ is denoted by $M_n(\mathbb{R})$. For the sake of simplicity, we shall only give proofs for the case $n = 2$. The results which we cover do generalize to the $n \times n$ case. Details can be found in a text on Linear Algebra, such as the book in this series by Allenby.

To define matrix multiplication, we first define $AB$ for a $1 \times n$ row matrix $A$ and an $n \times 1$ column matrix $B$. Notice that both $A$ and $B$ have $n$ entries.

$$\text{If } A = \begin{pmatrix} a_1 & a_2 & \ldots & a_n \end{pmatrix} \text{ and } B = \begin{pmatrix} b_1 \\ b_2 \\ \vdots \\ b_n \end{pmatrix} \text{ then } AB = (a_1b_1 + a_2b_2 + \ldots + a_nb_n).$$

For example,

$$\begin{pmatrix} 1 & -1 & 2 \end{pmatrix} \begin{pmatrix} -2 \\ 1 \\ 3 \end{pmatrix} = (-2 - 1 + 6) = (3).$$

In general, we only define a matrix product $AB$ when $A$ is $m \times n$ and $B$ is $n \times q$, that is $B$ has the same number of rows as $A$ has columns. Each row of $A$ and each column of $B$ have $n$ entries. We can now define $AB$ to be the $m \times q$ matrix where the entry in the $i$th row and $j$th column is found by multiplying the $i$th row of $A$ by the $j$th column of $B$. For example,

$$\begin{pmatrix} 1 & -1 & 2 \\ 2 & 1 & -1 \end{pmatrix} \begin{pmatrix} -2 & 1 \\ 1 & 0 \\ 3 & -1 \end{pmatrix} = \begin{pmatrix} 3 & -1 \\ -6 & 3 \end{pmatrix}.$$

The next exercise illustrates some of the basic properties of matrix multiplication.

## EXERCISE 1

(i) For each of the following pairs of matrices find the products $AB$ and $BA$.

(a) $A = \begin{pmatrix} 1 & 2 \\ -1 & 3 \end{pmatrix}, \quad B = \begin{pmatrix} 2 & 1 \\ -1 & 1 \end{pmatrix}.$

(b) $A = \begin{pmatrix} 2 & 4 \\ 1 & 2 \end{pmatrix}, \quad B = \begin{pmatrix} -2 & 4 \\ 1 & -2 \end{pmatrix}.$

(ii) Let $A = \begin{pmatrix} 2 & 0 \\ 0 & 2 \end{pmatrix}$, $B = \begin{pmatrix} 1 & 1 \\ 0 & 1 \end{pmatrix}$ and $X = \begin{pmatrix} x \\ y \end{pmatrix}$. Find $BX$, $A(BX)$, $AB$ and $(AB)X$.

(iii) Let $A = \begin{pmatrix} a & b \\ c & d \end{pmatrix}$, $B = \begin{pmatrix} e & f \\ g & h \end{pmatrix}$ and $C = \begin{pmatrix} p & q \\ r & s \end{pmatrix}$. Find $AB$, $(AB)C$, $BC$ and $A(BC)$.

(iv) Let $A = \begin{pmatrix} a & b \\ c & d \end{pmatrix}$ and $I = \begin{pmatrix} 1 & 0 \\ 0 & 1 \end{pmatrix}$. Show that $AI = A = IA$.

We list below some general observations based on Exercise 1.

- **Closure**. For each positive integer $n$, the set $M_n(\mathbb{R})$ is closed under matrix multiplication since multiplying two $n \times n$ matrices together results in an $n \times n$ matrix.
- **Associativity**. Matrix multiplication is associative, that is $(AB)C = A(BC)$ provided the products are defined. We will not prove this in general but Exercise 1(iii) checks the result when $A$, $B$ and $C$ are all $2 \times 2$. Since matrix multiplication is associative we have no need to include brackets when writing products of matrices.
- **Identity**. The matrix $I_2 = \begin{pmatrix} 1 & 0 \\ 0 & 1 \end{pmatrix}$ is called the $2 \times 2$ **identity** matrix. In Exercise 1(iv), $AI_2 = I_2A = A$ for all $2 \times 2$ matrices $A$. In other words $I_2$ is neutral for matrix multiplication in $M_2(\mathbb{R})$. In general, the $n \times n$ identity matrix $I_n$, which is the $n \times n$ matrix

$$\begin{pmatrix} 1 & 0 & \dots & 0 \\ 0 & 1 & \dots & 0 \\ \vdots & \vdots & \ddots & \vdots \\ 0 & 0 & \dots & 1 \end{pmatrix},$$

is neutral for matrix multiplication in $M_n(\mathbb{R})$.

- **Zero matrix**. The matrix $\begin{pmatrix} 0 & 0 \\ 0 & 0 \end{pmatrix}$ is known as the $2 \times 2$ zero matrix. Notice, from Exercise 1(i)(b), that it can be the result of multiplying two non-zero matrices together.
- **Non-commutativity.** Let $A, B \in M_2(\mathbb{R})$. As can be seen from Exercise 1(i)(a), it is possible that $AB \neq BA$.

In previous chapters, when we have had a neutral element we proceeded to discuss inverses. When does a matrix have an inverse? An $n \times n$ matrix $A$ is **invertible** with **inverse** $B$, if there is an $n \times n$ matrix $B$ such that $AB = BA = I_n$. We shall show in a more general context later that $A$ cannot have more than one inverse. Thus, we can write the inverse of an invertible matrix $A$ as $A^{-1}$.

If $A = \begin{pmatrix} a & b \\ c & d \end{pmatrix}$ and $ad - bc \neq 0$ then it can be easily checked that $A$ has inverse

$$A^{-1} = \begin{pmatrix} \frac{d}{ad-bc} & \frac{-b}{ad-bc} \\ \frac{-c}{ad-bc} & \frac{a}{ad-bc} \end{pmatrix}.$$

Now suppose that $ad - bc = 0$. Can $A$ have an inverse in this case? Let $B = \begin{pmatrix} d & -b \\ -c & a \end{pmatrix}$. Then

$$AB = \begin{pmatrix} a & b \\ c & d \end{pmatrix}\begin{pmatrix} d & -b \\ -c & a \end{pmatrix} = \begin{pmatrix} ad-bc & 0 \\ 0 & ad-bc \end{pmatrix} = \begin{pmatrix} 0 & 0 \\ 0 & 0 \end{pmatrix}.$$

Suppose that there is a matrix $X$ such that $XA = I_2$. Then

$$B = I_2 B = (XA)B = X(AB) = X0 = 0.$$

Hence $a = b = c = d = 0$ and so $A = 0$. This would give $I_2 = XA = 0$ which is impossible. Thus, if $ad - bc = 0$ then the matrix $A$ does not have an inverse.

The number $ad - bc$ is clearly significant here and is known as the **determinant** of $A$ and written $\det A$. The idea of a determinant generalizes to $n \times n$ matrices. The general definition and properties can be found in the book in this series by Allenby.

## Theorem 1

Let $A, B \in M_n(\mathbb{R})$. Then

(i) $A$ is invertible if and only if $\det A \neq 0$.

(ii) $\det A \det B = \det(AB)$.

(iii) If $A$ and $B$ are invertible matrices then $AB$ is invertible. Moreover $(AB)^{-1} = B^{-1}A^{-1}$.

PROOF

(for $n = 2$.)

(i) Above we have shown that if $\det A \neq 0$ then $A$ has an inverse. Also we have checked that if $\det A = 0$ then $A$ does not have an inverse, in other words, $A$ has an inverse only if $\det A \neq 0$. Together these give the result for $n = 2$.

(ii) Let $A = \begin{pmatrix} a & b \\ c & d \end{pmatrix}$ and $B = \begin{pmatrix} e & f \\ g & h \end{pmatrix}$. Then

$$\det A \det B = (ad - bc)(eh - fg) = adeh + bcfg - bceh - adfg.$$

$$\text{Now } AB = \begin{pmatrix} ae + bg & af + bh \\ ce + dg & cf + dh \end{pmatrix}$$

and so

$$\det(AB) = (ae + bg)(cf + dh) - (af + bh)(ce + dg) = aedh + bgcf - bhce - afdg.$$

(iii) Suppose that $A$ and $B$ are invertible. Then $(B^{-1}A^{-1})(AB) = B^{-1}I_2B = I_2$ and similarly $(AB)(B^{-1}A^{-1}) = I_2$. Thus, $AB$ is invertible with $(AB)^{-1} = B^{-1}A^{-1}$ as required. Note that we have seen a similar formula for the inverse of the composition of two invertible functions in Chapter 2. ●

### Group properties for invertible matrices

We are now in a position to see that, under matrix multiplication, the set of invertible $n \times n$ real matrices satisfies the four group properties described on page 5. This set will be called the **general linear group**, $GL_n(\mathbb{R})$.

- **Closure** of $GL_n(\mathbb{R})$ is shown in Theorem 1(iii): if $A$ and $B$ are invertible then $AB$ is invertible.
- Matrix multiplication is **associative**.
- There is a **neutral** element namely the identity matrix $I_n$, which is certainly invertible, being its own inverse.
- If $A \in GL_n(\mathbb{R})$ has inverse $B$ then $B$ has inverse $A$ so $B \in GL_n(\mathbb{R})$. Thus, we have the existence of **inverses** in $GL_n(\mathbb{R})$.

**EXERCISE 2**

For each of the following $2 \times 2$ matrices $A$, find, if possible, the inverse of $A$.

(i) $A = \begin{pmatrix} 1 & 1 \\ 1 & 2 \end{pmatrix}$;

(ii) $A = \begin{pmatrix} -1 & 1 \\ -4 & 2 \end{pmatrix}$;

(iii) $A = \begin{pmatrix} 2 & 6 \\ 1 & 3 \end{pmatrix}$.

## 3.2 Linear Transformations

How can we use a $2 \times 2$ matrix to obtain a function from $\mathbb{R}^2$ to itself? It will be convenient to write points of $\mathbb{R}^2$ as $2 \times 1$ matrices. Thus, we write $\begin{pmatrix} x \\ y \end{pmatrix}$ rather than $(x, y)$. If $v = \begin{pmatrix} x \\ y \end{pmatrix}$ and $A = \begin{pmatrix} a & b \\ c & d \end{pmatrix} \in M_2(\mathbb{R})$ then $Av$ is another $2 \times 1$ matrix, that is $Av \in \mathbb{R}^2$. Thus, we have a function $f_A : \mathbb{R}^2 \to \mathbb{R}^2$ given by $f_A(v) = Av$, that is

$$f_A : \begin{pmatrix} x \\ y \end{pmatrix} \mapsto \begin{pmatrix} a & b \\ c & d \end{pmatrix}\begin{pmatrix} x \\ y \end{pmatrix} = \begin{pmatrix} ax + by \\ cx + dy \end{pmatrix}.$$

The function $f_A$ is called the **linear transformation** of $\mathbb{R}^2$ determined by $A$.

## Example 1

If $A = \begin{pmatrix} 2 & 0 \\ 0 & 2 \end{pmatrix}$ then $f_A$ sends $\begin{pmatrix} x \\ y \end{pmatrix}$ to $\begin{pmatrix} 2x \\ 2y \end{pmatrix}$. Here, $f_A$ is the enlargement shown in Fig 3.1.

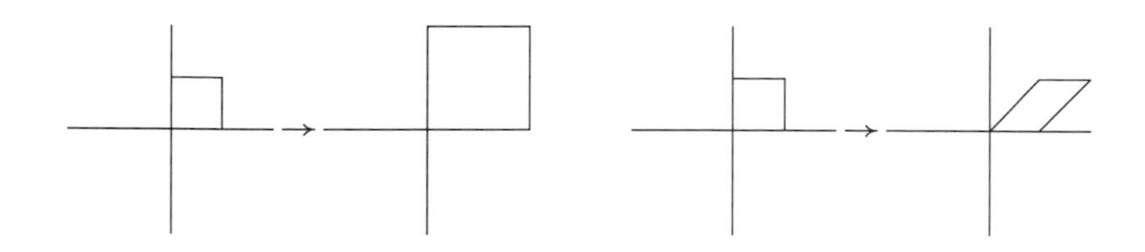

**Fig 3.1** An enlargement and a shear.

## Example 2

Let $a \in \mathbb{R}$ and let

$$B = \begin{pmatrix} 1 & a \\ 0 & 1 \end{pmatrix}, \quad f_B : \begin{pmatrix} x \\ y \end{pmatrix} \mapsto \begin{pmatrix} x + ay \\ y \end{pmatrix}.$$

When $a = 1$ this is shown in Fig 3.1 and, in general, it is a **shear** parallel to the $x$-axis. Each point $\begin{pmatrix} x \\ y \end{pmatrix}$ is moved parallel to the $x$-axis by a distance proportional to $y$.

## Example 3

$$C = \begin{pmatrix} 0 & 0 \\ 0 & 1 \end{pmatrix}, \quad f_C : \begin{pmatrix} x \\ y \end{pmatrix} \mapsto \begin{pmatrix} 0 \\ y \end{pmatrix}.$$

This transformation is called **projection** onto the $y$-axis. Notice that the matrix $C$ is not invertible and that the function $f_C$ is neither surjective nor injective and so is not invertible.

## EXERCISE 3

Let $A$ and $B$ be the matrices in the first two examples above. Show that $f_A \circ f_B = f_{AB}$.

This exercise illustrates how the composition of transformations corresponds to matrix multiplication. Since this is an important instance of the idea of homomorphism, which we shall meet later, we prove the general result in the next theorem.

## Theorem 2

Let $A$ and $B$ be $2 \times 2$ matrices.

(i) $f_A \circ f_B = f_{AB}$.

(ii) If $A$ is invertible then $f_A$ is an invertible function with inverse $f_{A^{-1}}$. ●

PROOF

(i) Both $f_A \circ f_B$ and $f_{AB}$ are functions from $\mathbb{R}^2$ to itself. Let $v \in \mathbb{R}^2$. Then, by associativity of matrix multiplication,

$$(f_A \circ f_B)(v) = f_A(f_B(v)) = f_A(Bv) = A(Bv) = (AB)v = f_{AB}(v).$$

Thus $f_A \circ f_B = f_{AB}$.

(ii) Now $f_{I_2}$ is the identity function $\mathrm{id}_{\mathbb{R}^2}$ since $I_2 v = v$ for all $v \in \mathbb{R}^2$. By the first part, $f_A \circ f_{A^{-1}} = f_{AA^{-1}} = f_{I_2} = \mathrm{id}_{\mathbb{R}^2}$. Similarly $f_{A^{-1}} \circ f_A = \mathrm{id}_{\mathbb{R}^2}$. Hence, $f_A$ has inverse $f_{A^{-1}}$. ●

**EXERCISE 4**

For each of the following matrices $A$, find a formula for $f_A\left(\begin{pmatrix} x \\ y \end{pmatrix}\right)$ and describe the effect of $f_A$ geometrically:

$$\begin{pmatrix} 1 & 0 \\ 2 & 1 \end{pmatrix}; \quad \begin{pmatrix} -1 & 0 \\ 0 & -1 \end{pmatrix}; \quad \begin{pmatrix} 0 & 1 \\ 1 & 0 \end{pmatrix}.$$

## 3.3 Orthogonal Matrices

In Chapter 1 we met the functions $\mathrm{rot}_\phi$ and $\mathrm{ref}_\phi$. These are, in fact, linear transformations. To see this, let $A_\phi = \begin{pmatrix} \cos\phi & -\sin\phi \\ \sin\phi & \cos\phi \end{pmatrix}$, $B_\phi = \begin{pmatrix} \cos\phi & \sin\phi \\ \sin\phi & -\cos\phi \end{pmatrix}$ and $v = \begin{pmatrix} r\cos\theta \\ r\sin\theta \end{pmatrix}$. Using standard trigonometric identities, it is routine to check that $A_\phi v = \mathrm{rot}_\phi(v)$ and $B_\phi v = \mathrm{ref}_\phi(v)$. This shows that $f_{A_\phi} = \mathrm{rot}_\phi$ and $f_{B_\phi} = \mathrm{ref}_\phi$.

### Transpose of a matrix

If we reflect a square matrix $A$ in its main diagonal (top left to bottom right) we obtain another square matrix called the **transpose** of $A$. For example, the transpose of $\begin{pmatrix} a & b \\ c & d \end{pmatrix}$ is $\begin{pmatrix} a & c \\ b & d \end{pmatrix}$. We denote the transpose of $A$ by $A^T$. Formally, we can define $A^T$ by saying that the $ij$th entry of $A^T$ is the $ji$th entry of $A$.

**EXERCISE 5**

(i) Let $A = \begin{pmatrix} 1 & 2 & 6 \\ 2 & -1 & 1 \\ 3 & 7 & 2 \end{pmatrix}$. Write down $A^T$.

(ii) How can the expression $(A^T)^T$ be simplified?

## • Theorem 3

If $A$ and $B$ are $n \times n$ matrices then $(AB)^T = B^T A^T$ and $\det(A^T) = \det A$. ●

PROOF

For the case $n = 2$, these are straightforward calculations which we leave as exercises. ●

Note the reversal of order here. The theorem may remind you of a similar result for inverses of matrices. There exist matrices $A$ with $A^T = A^{-1}$. Such matrices are called **orthogonal** and the set of all orthogonal $2 \times 2$ matrices is denoted by $O_2$. The reason that this is the same notation as for the orthogonal group will become apparent in Chapter 9.

As examples of orthogonal matrices, we have the rotation matrix $A_\phi = \begin{pmatrix} \cos\phi & -\sin\phi \\ \sin\phi & \cos\phi \end{pmatrix}$ and the reflection matrix $B_\phi = \begin{pmatrix} \cos\phi & \sin\phi \\ \sin\phi & -\cos\phi \end{pmatrix}$. It is easy to check, using the formula for inverse on page 31, that $A_\phi$ and $B_\phi$ are orthogonal. The next theorem shows that they are the only $2 \times 2$ orthogonal matrices.

## • Theorem 4

Let $A \in O_2$ be an orthogonal matrix. Then there is $\phi \in \mathbb{R}$ with $0 \leqslant \phi < 2\pi$ such that $A$ is either the rotation matrix $A_\phi$ or the reflection matrix $B_\phi$. ●

PROOF

Let $A = \begin{pmatrix} a & b \\ c & d \end{pmatrix}$ so that $A^T = \begin{pmatrix} a & c \\ b & d \end{pmatrix}$. Using Theorems 3 and 1(ii),

$$(\det A)^2 = \det A \det A^T = \det(AA^T) = \det I = 1.$$

Hence, $\det A = \pm 1$. Firstly, suppose that $\det A = 1$. Then $A^T = A^{-1} = \begin{pmatrix} d & -b \\ -c & a \end{pmatrix}$, so $d = a$ and $b = -c$. Hence, $A = \begin{pmatrix} a & -c \\ c & a \end{pmatrix}$ where $a^2 + c^2 = 1$. Now suppose that $\det A = -1$. In this case $A^T = A^{-1} = \begin{pmatrix} -d & b \\ c & -a \end{pmatrix}$, so $d = -a$ and $b = c$. Hence, $A = \begin{pmatrix} a & c \\ c & -a \end{pmatrix}$, where $-a^2 - c^2 = -1$. In both cases, $a^2 + c^2 = 1$, so there is an angle $\phi$, with $0 \leqslant \phi < 2\pi$, such that $a = \cos\phi$ and $c = \sin\phi$, see Fig 3.2.

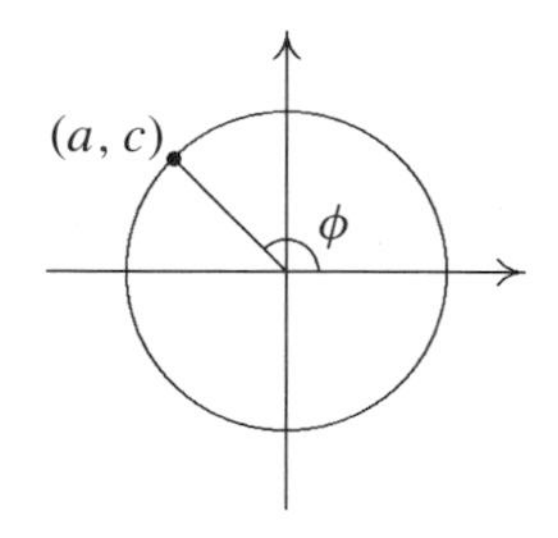


**Fig 3.2** Choosing $\phi$ from $a$ and $c$.

Hence $A = \begin{pmatrix} \cos\phi & -\sin\phi \\ \sin\phi & \cos\phi \end{pmatrix} = A_\phi$ or $A = \begin{pmatrix} \cos\phi & \sin\phi \\ \sin\phi & -\cos\phi \end{pmatrix} = B_\phi$. ●

**EXERCISE 6**

(i) Is the set of orthogonal matrices closed under multiplication?

(ii) Is the inverse of an orthogonal matrix always orthogonal?

## 3.4 Further Exercises on Chapter 3

**EXERCISE 7**

Let $A = \begin{pmatrix} 1 & 2 \\ 0 & 1 \end{pmatrix}$. Find all $2 \times 2$ matrices $B$ such that $AB = BA$.

**EXERCISE 8**

Let $A$ be an invertible matrix and $B$ be a matrix such that $AB = BA$. Show that $A^{-1}B = BA^{-1}$.

**EXERCISE 9**

A shear parallel to the $x$-axis is a linear transformation determined by a matrix $A$ with $A = \begin{pmatrix} 1 & a \\ 0 & 1 \end{pmatrix}$, where $a \in \mathbb{R}$. An example of a shear is shown in Fig 3.1.

(i) Is the set of shears closed under composition?

(ii) Is the inverse of a shear always a shear?

**EXERCISE 10**

A $2 \times 2$ **diagonal** matrix has the form $D = \begin{pmatrix} b & 0 \\ 0 & c \end{pmatrix}$.

(i) Which diagonal matrices are invertible?

(ii) Is the set of invertible diagonal matrices closed under multiplication?

(iii) Is the inverse of an invertible diagonal matrix always diagonal?

(iv) Describe the effect of the linear transformation determined by $D$.

**EXERCISE 11**

A $2 \times 2$ **upper triangular** matrix has the form $U = \begin{pmatrix} b & d \\ 0 & c \end{pmatrix}$.

(i) Which upper triangular matrices are invertible?

(ii) Is the set of invertible upper triangular matrices closed under multiplication?

(iii) Is the inverse of an invertible upper triangular matrix always upper triangular?

(iv) Show that every invertible upper triangular matrix can be written as a product $AB$ where $A$ is the matrix of a shear parallel to the $x$-axis and $B$ is an invertible diagonal matrix.

## EXERCISE 12

A $2 \times 2$ matrix $A$ is **symmetric** if $A = A^T$. Is the set of invertible symmetric $2 \times 2$ matrices closed under multiplication?

## EXERCISE 13

Find a general formula for a product $P = AB$ where $A$ is the matrix of a shear parallel to the $x$-axis and $B$ is the matrix of a shear parallel to the $y$-axis. Is the set of all such matrices $P$ closed under multiplication? Investigate the corresponding problem for other products of matrices of different types (for example, diagonal, rotation, reflection, shears, upper triangular).

# 4 • The Group Axioms

In this chapter we take the group properties which we have observed in symmetries, permutations and invertible matrices and formalize them in the definition of a group. This reflects the historical development of the subject. In the early nineteenth century, the study of equations led to groups of permutations (or substitutions as they were then called). Later, groups of transformations were studied and common features were observed. It was not until the end of the century that some of these features were abstracted as axioms and abstract groups were defined.

## 4.1 Number Systems

The examples of groups which we have discussed so far are groups of functions or matrices. Other examples come from familiar number systems. This section covers the basic notation and properties of these. We have already met the notation $\mathbb{R}$ for the set of real numbers. Other important sets of numbers are:

- $\mathbb{N} = \{1, 2, 3, \ldots\}$, the set of **natural numbers** or **positive integers** (some texts include 0 in $\mathbb{N}$ but we shall not do so);
- $\mathbb{Z} = \{\ldots, -3, -2, -1, 0, 1, 2, 3, \ldots\}$, the set of **integers** (the letter $\mathbb{Z}$ comes from the German word *zahlen* for numbers);
- $\mathbb{Q}$, the set of **rational numbers** $\frac{p}{q}$ where $p$ and $q$ are integers with $q \neq 0$ ($\mathbb{Q}$ comes from *quotient*);
- $\mathbb{C}$, the set of **complex numbers**.

Most readers will be familiar with complex numbers, which are numbers of the form $a + b\mathrm{i}$ where $a, b \in \mathbb{R}$ and $\mathrm{i}^2 = -1$. Elements are added and multiplied together according to the rules

$$(a + b\mathrm{i}) + (c + d\mathrm{i}) = (a + c) + (b + d)\mathrm{i} \text{ and } (a + b\mathrm{i})(c + d\mathrm{i}) = (ac - bd) + (ad + bc)\mathrm{i}.$$

Note that $b$ may take the value 0 and so $\mathbb{R} \subset \mathbb{C}$. Whereas the equation $x^2 + 1$ has no solution in $\mathbb{R}$, it has two solutions in $\mathbb{C}$, namely i and $-$i. Indeed, any polynomial equation of degree $n$ with coefficients in $\mathbb{R}$, or even in $\mathbb{C}$, has $n$ solutions in $\mathbb{C}$. This is the *Fundamental Theorem of Algebra* and its proof is beyond the scope of this book.

We list below some basic properties of complex numbers.

- Every complex number $z = a + b\mathrm{i}$ can be represented as a point $(a, b)$ in $\mathbb{R}^2$. If the polar coordinates of $(a, b)$ are $r$ and $\theta$ then $a = r\cos\theta$, $b = r\sin\theta$ and $z = r(\cos\theta + \mathrm{i}\sin\theta)$. This is called the **polar form** of $z$ and it is sometimes abbreviated to $r\mathrm{e}^{\mathrm{i}\theta}$.
- $\mathrm{e}^{\mathrm{i}\pi} = -1$ and $\mathrm{e}^{2n\mathrm{i}\pi} = 1$ if $n \in \mathbb{Z}$.

- In polar form, the rule for multiplication is $re^{i\theta}se^{i\phi} = rse^{i(\theta+\phi)}$. In particular, when $s = r$ and $\phi = \theta$, $(re^{i\theta})^2 = r^2e^{2i\theta}$.

- The above rule for squares extends to all positive integers $n$: $(re^{i\theta})^n = r^ne^{in\theta}$. This is known as De Moivre's Theorem.

- There are $n$ distinct $n$th roots of 1, namely $e^{\frac{2ki\pi}{n}}$ for each $k = 0, 1 \ldots, n-1$; if $z = e^{\frac{2ki\pi}{n}}$ then $z^n = e^{2ki\pi} = 1$. For example, the four fourth roots of 1 are $1$, $e^{\frac{i\pi}{2}} = i$, $e^{i\pi} = -1$ and $e^{\frac{3i\pi}{4}} = -i$. These are the four roots of the equation $x^4 - 1 = 0$.

## Prime numbers and factorization

We shall often need to count the number of elements in a group. In this context prime numbers and factorization are significant and we list below those aspects which we shall need. More details may be found in the book by Allenby and Redfern.

- A positive integer $p \neq 1$ is called **prime** if it has no positive integer factors apart from itself and 1. The first eight prime numbers are 2, 3, 5, 7, 11, 13, 17 and 19.

- Every positive integer can be expressed as a product $p_1^{a_1}p_2^{a_2}\ldots p_k^{a_k}$ of powers of primes in a unique way. For example $126 = 2 \times 3^2 \times 7$ and $120 = 2^3 \times 3 \times 5$. This is the *Fundamental Theorem of Arithmetic.*

Let $m$ and $n$ be two positive integers and let $p_1, p_2, \ldots p_k$ be the prime numbers occurring in either of their prime factorizations. Then $m = p_1^{a_1}p_2^{a_2}\ldots p_k^{a_k}$ and $n = p_1^{b_1}p_2^{b_2}\ldots p_k^{b_k}$ where $a_i$ or $b_i$, but not both, might be zero. For example, if $m = 126$ and $n = 120$ then $p_1 = 2$, $p_2 = 3$, $p_3 = 5$, $p_4 = 7$ and $126 = 2 \times 3^2 \times 5^0 \times 7$ and $120 = 2^3 \times 3 \times 5 \times 7^0$.

- The prime factorization of $mn$ is $p_1^{a_1+b_1}p_2^{a_2+b_2}\ldots p_k^{a_k+b_k}$ so if $p$ is a prime factor of $mn$ then $p$ must be a prime factor of either $m$ or $n$ or both.

- The **highest common factor** or **h.c.f** of $m$ and $n$ is the largest positive integer $h$ that is a factor of both. If $c_i$ is the smaller of $a_i$ and $b_i$ for each $i$, then $h$ has prime factorization $p_1^{c_1}p_2^{c_2}\ldots p_k^{c_k}$. For example, the h.c.f. of 126 and 120 is $2^13^15^07^0 = 2 \times 3 = 6$.

- The **least common multiple** or **l.c.m.** of $m$ and $n$ is the smallest positive integer $l$ that is a multiple of both $m$ and $n$. If $d_i$ is the larger of $a_i$ and $b_i$ for each $i$, then $l$ has prime factorization $p_1^{d_1}p_2^{d_2}\ldots p_k^{d_k}$. For example, the l.c.m of 126 and 120 is $2^3 \times 3^2 \times 5^1 \times 7^1 = 2520$.

- Because $c_i + d_i$ is always $a_i + b_i$, we can see that $mn = hl$. Thus, for 126 and 120, $126 \times 120 = 6 \times 2520 = 15120$.

- Two positive integers $m$ and $n$ are said to be **coprime** if their highest common factor is 1. In this case their l.c.m is $mn$.

# 4.2 Binary Operations

The first ingredient for a group is a method of combining elements of a set together. Let $A$ be a non-empty set. A **binary operation** $\odot$ on $A$ is a rule which, for each ordered pair $(a, b)$ of elements of $A$, determines a unique element $a \odot b$ of $A$. The word 'binary' is used since just two elements are combined by the operation. When $\odot$ is a binary operation on $A$ we often say that $A$ is **closed** under $\odot$ and refer to the **closure** of $\odot$ on $A$.

The word 'ordered' in this definition is important; $a \odot b$ need not be the same as $b \odot a$.

## Examples of binary operations

(i) Addition, $+$, is a binary operation on each of the sets $\mathbb{N}$, $\mathbb{Z}$, $\mathbb{Q}$, $\mathbb{R}$ and $\mathbb{C}$ since adding any two numbers from one of these sets together gives another number in the set.

(ii) Subtraction, $-$, is a binary operation on each of the sets $\mathbb{Z}$, $\mathbb{Q}$, $\mathbb{R}$ and $\mathbb{C}$ but it is **not** a binary operation on $\mathbb{N}$ since, for example, although both 1 and 2 belong to $\mathbb{N}$, $1 - 2 = -1 \notin \mathbb{N}$. Notice that just one example like this is sufficient to show that a rule is not a binary operation on a given set.

(iii) Multiplication, $\times$, is a binary operation on each of the sets $\mathbb{N}$, $\mathbb{Z}$, $\mathbb{Q}$, $\mathbb{R}$ and $\mathbb{C}$.

(iv) Division, $\div$, is not a binary operation on any of sets $\mathbb{Z}$, $\mathbb{Q}$, $\mathbb{R}$ and $\mathbb{C}$ since for any number $a$, $a \div 0$ is not defined. However, $\div$ is a binary operation on each of the sets $\mathbb{Q}\backslash\{0\}$, $\mathbb{R}\backslash\{0\}$ and $\mathbb{C}\backslash\{0\}$ of non-zero numbers. The same is not true of $\mathbb{Z}\backslash\{0\}$ as $1 \in \mathbb{Z}\backslash\{0\}$ and $2 \in \mathbb{Z}\backslash\{0\}$ but $\frac{1}{2} \notin \mathbb{Z}\backslash\{0\}$.

(v) For a given non-empty set $X$, composition of functions, $\circ$, is a binary operation on the set of all permutations of $X$ because, as we observed in Chapter 2, the composition of two permutations is a permutation.

(vi) Sometimes we only wish to consider some of the permutations of $X$. This was so in Chapter 1 where $X = \mathbb{R}^2$ and we considered the symmetries of the square or those of the circle. In each of these cases the set of symmetries is closed under composition. In the case of the square this can be seen from the Cayley Table 1.2 and for the circle it follows from Table 1.1. Thus, composition is a binary operation on each of these sets of symmetries.

(vii) Matrix multiplication is a binary operation on the set of all $2 \times 2$ real matrices and, because the product of any two invertible $2 \times 2$ matrices is invertible, on the set of all $2 \times 2$ invertible real matrices.

## EXERCISE 1

(i) Let $A$ be the set of negative real numbers. Is addition a binary operation on $A$? Is multiplication a binary operation on $A$? Is subtraction a binary operation on $A$?

(ii) There is a binary operation $\dagger$ on $\mathbb{R}$ given by the rule $a \dagger b = ab + a + b$. Show that $\mathbb{R}\backslash\{-1\}$ is closed under $\dagger$, that is, that if $a \neq -1$ and $b \neq -1$ then $a \dagger b \neq -1$.

# 4.3 Definition of a Group

We say that a non-empty set $G$ is a **group** under $\odot$, or, more formally, that $(G, \odot)$ is a group, if the following four axioms hold.

### G1 [Closure]:

$\odot$ is a binary operation on $G$, that is $a \odot b \in G$ for all $a, b \in G$.

### G2 [Associativity]:

$(a \odot b) \odot c = a \odot (b \odot c)$ for all $a, b, c \in G$.

### G3 [Existence of identity]:

there is an element $e \in G$ with the property that $e \odot g = g \odot e = g$ for all $g \in G$. We call such an element $e$ an **identity** or **neutral** element of $G$.

### G4 [Existence of inverses]:

for each element $g \in G$ there is an element $h \in G$ with the property that $g \odot h = h \odot g = e$. We call such an element $h$ an **inverse** of $g$.

The definition of group $G$ does not require that $a \odot b = b \odot a$ for all $a, b \in G$. We shall see that some basic examples have this extra property but others have not. We say that a group $(G, \odot)$ is **abelian** or **commutative** if $a \odot b = b \odot a$ for every $a, b \in G$. If $a, b \in G$ are such that $a \odot b = b \odot a$ we say that $a$ **commutes** with $b$ or that $a$ and $b$ **commute**.

One measure of a group, indeed of any set $A$, is the number of elements in $A$. We shall write this as $|A|$. For a group $G$, we call $|G|$ the **order** of $G$. Thus, the order of $G$ is either a positive integer or is infinite. In the latter case we write $|G| = \infty$.

# 4.4 Examples of Groups

## Groups of numbers under addition

The familiar sets $\mathbb{C}$, $\mathbb{R}$, $\mathbb{Q}$ and $\mathbb{Z}$ are all groups under addition. They are closed under addition, which is associative. The neutral element in each case is 0 and the inverse of $a$ is $-a$. For example, $2 + (-2) = 0 = (-2) + 2$. These groups are all abelian and of infinite order.

The sets $\mathbb{N} = \{1, 2, 3, \ldots\}$ and $\{0, 1, 2, 3, \ldots\}$ are not groups under addition. The former has no neutral element. The latter does have a neutral element, namely 0, but only 0 has an inverse in $\{0, 1, 2, 3, \ldots\}$.

## Groups of numbers under multiplication

The sets $\mathbb{C}$, $\mathbb{R}$, $\mathbb{Q}$ and $\mathbb{Z}$ are not groups under multiplication. The axioms G1,G2 and G3 all hold, with neutral element 1, but 0 has no inverse. However, if $X = \mathbb{C}$, $\mathbb{R}$ or $\mathbb{Q}$ then $(X\backslash\{0\}, \times)$ is a group. The neutral element is 1 and the inverse of $x$ is $\frac{1}{x}$ or $x^{-1}$. These groups are all abelian and of infinite order.

On the other hand, $(\mathbb{Z}\backslash\{0\}, \times)$ is not a group since only two elements, 1 and $-1$, have inverses in $\mathbb{Z}\backslash\{0\}$. For example, the only possible inverse for 2 under multiplication is $\frac{1}{2}$ which is not in $\mathbb{Z}\backslash\{0\}$. If we just take those elements of $\mathbb{Z}$ which have multiplicative inverses in $\mathbb{Z}$, we obtain $\{1, -1\}$. This is clearly closed under multiplication and is a group in which each element is its own inverse. This group is abelian of order 2.

## Groups of matrices

Fix a positive integer $n$. Under matrix multiplication, $n \times n$ real matrices do not form a group since not all $n \times n$ matrices have inverses. However, if we take only those $n \times n$ real matrices which do have inverses we obtain the **general linear group over** $\mathbb{R}$, denoted by $GL_n(\mathbb{R})$. The axioms for this group were checked in Chapter 3 for the case $n = 2$. Similarly, we get general linear groups $GL_n(\mathbb{Q})$ and $GL_n(\mathbb{C})$ over $\mathbb{Q}$ and $\mathbb{C}$. These groups all have infinite order. Unless $n = 1$, they are not abelian.

## Groups of functions under composition

We have already met several examples of groups of functions under composition and have checked that composition is associative. In each of these groups the identity function is neutral.

- The permutations of a set $X$ form a group under composition. This was checked in Chapter 2 where this group was denoted by $S_X$. In particular, when $X = \{1, 2, \ldots, n\}$ we obtain the **symmetric group** $S_n$. We have seen in Chapter 2 that $S_n$ has $n!$ elements. Thus, $|S_n| = n!$. If $|X| = \infty$ then $|S_X| = \infty$. If $n > 2$ then $S_n$ is not abelian. To show this, it is enough to find two elements with $fg \neq gf$. For example, $(1\ 2)(1\ 3) \neq (1\ 3)(1\ 2)$. Of course it is not true that $fg \neq gf$ for all $f, g \in S_n$. We have seen that disjoint cycles commute, for example $(1\ 2)(3\ 4) = (3\ 4)(1\ 2)$.
- We looked at symmetries of squares and circles in Chapter 1. The group axioms G1, G3 and G4 were checked, under composition, for the symmetries of the square and for the symmetries of the circle. The other axiom G2, associativity, always holds for composition. Thus, we obtain two groups, the **dihedral group**, $D_4$, of symmetries of the square and the **orthogonal group**, $O_2$, of symmetries of the circle. These have order 8 and $\infty$ respectively. They are not abelian because they both contain $\text{rot}_{\frac{\pi}{2}}$ and $\text{ref}_{\frac{\pi}{2}}$. You should check as an exercise that these do not commute.
- As we shall see later, symmetries of a figure always form a group under composition. For now we shall just consider one more such group, namely the symmetries of a non-square rectangle. These are shown in Fig 4.1.

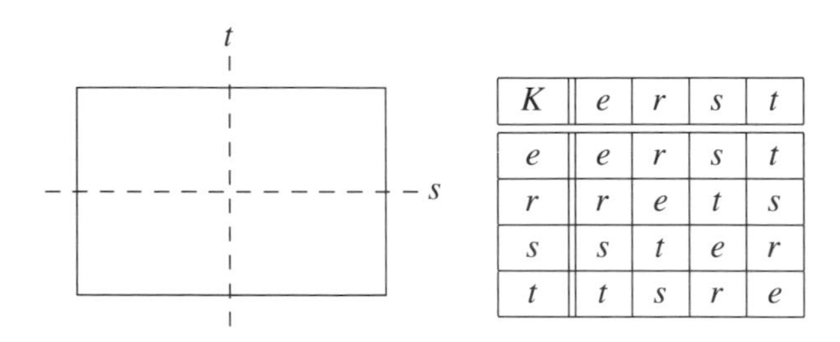


| $K$ | $e$ | $r$ | $s$ | $t$ |
|---|---|---|---|---|
| $e$ | $e$ | $r$ | $s$ | $t$ |
| $r$ | $r$ | $e$ | $t$ | $s$ |
| $s$ | $s$ | $t$ | $e$ | $r$ |
| $t$ | $t$ | $s$ | $r$ | $e$ |

**Fig 4.1** Symmetries of a rectangle.

This group, $K$, say, has four elements $e = \text{rot}_0$, $r = \text{rot}_\pi$, $s = \text{ref}_0$ and $t = \text{ref}_\pi$. The Cayley table shows that it is closed with $e$ as the neutral element, that each element is its own inverse and that the group is abelian.

- Consider the set $G$ of all linear transformations of $\mathbb{R}^2$ of the form $f_A$ where $A$ is an invertible $2 \times 2$ real matrix. As $f_A \circ f_B = f_{AB}$ (by Theorem 2 of Chapter 3) and as invertible matrices are closed under multiplication, $G$ is closed under composition which is associative. The identity matrix determines the identity function $(f_I(v) = Iv = v)$ and each $f_A \in G$ has inverse $f_{A^{-1}} \in G$. Thus, $G$ is a group under composition.

### Notation

We shall often omit the symbol $\odot$ for a binary operation and simply use juxtaposition, that is we write $ab$ instead of $a \odot b$. In particular, we do this when the binary operation is multiplication (of numbers or matrices) or composition of functions. However, when the binary operation is addition we include the $+$ sign as juxtaposition would be confusing. In this case, we shall say that the group is in **additive notation.**

## 4.5 Consequences of the Axioms

The four axioms for a group represent common properties of the various basic examples of groups but are not the only common properties. Below, we note some other common properties of all groups. The reason these are not added to the list of axioms is that they can be deduced as consequences of the axioms.

### Uniqueness of identity

The axioms say that a group must have a neutral element, but could it have more than one? Suppose that the group $G$ has two neutral elements $e$ and $f$. Then $ef = e$ since $f$ is neutral, and $ef = f$ since $e$ is neutral. Thus, $f = e$. This shows that there cannot be two different neutral elements. In a specific example we often use the usual notation for the identity in that example, rather than $e$. Thus, we might write a neutral element as

- 1 in a group of numbers under multiplication,
- $I$ in a group of matrices under matrix multiplication,
- id in a group of functions under composition,
- 0 in a group in additive notation.

Sometimes we denote the neutral element of a group $G$ by $e_G$. This avoids confusion when there is more than one group around with possibly distinct neutral elements.

### Uniqueness of inverses

We also know from the axioms that each element $g$ of a group $G$ must have an inverse but can it have more than one? Suppose that $g$ has two inverses $h$ and $k$ in $G$. Then $h = he = h(gk) = (hg)k = ek = k$. Thus, the inverse of $g$ is unique. We denote the inverse of $g$ by $g^{-1}$. In additive notation, however, we denote the inverse of $g$ by $-g$.

## Omission of brackets

The associative law allows us to write $abc$ without ambiguity. Indeed, we can write a string of elements of any length, for example $ghkghggk$, without brackets. For a group in additive notation, we can write $a + b + c$ without brackets.

## Powers

Let $G$ be a group and let $g \in G$. Then $gg \in G$ and is written $g^2$. For an arbitrary positive integer $n$

$$g^n = \underbrace{gg \dots g}_{n \text{ times}}.$$

For positive integers $m$ and $n$, we have the familiar laws for indices

$$g^m g^n = g^{m+n} \text{ and } (g^m)^n = g^{mn}.$$

We define

$$g^0 = e, \text{ and } g^{-n} = (g^{-1})^n.$$

The above laws for indices then extend to all integers $m$ and $n$.

When additive notation is being used, the notation $g^n$ for $g + g + \dots g$ would be confusing so we write

$$ng = \underbrace{g + g + \dots + g}_{n \text{ times}}, \quad 0g = 0 \text{ and } (-n)g = n(-g).$$

We then have the laws

$$mg + ng = (m + n)g \text{ and } m(ng) = (mn)g \text{ for all } m, n \in \mathbb{Z}.$$

## Inverses of products

Let $g, h$ be two elements of a group $G$. As $gh \in G$, it has an inverse in $G$. The formula for this is

$$(gh)^{-1} = h^{-1}g^{-1}.$$

We check this as follows:

$$\begin{aligned}(gh)h^{-1}g^{-1} &= g(hh^{-1})g^{-1} \text{ using the associative law,}\\ &= geg^{-1}\\ &= gg^{-1}\\ &= e.\end{aligned}$$

Similarly $h^{-1}g^{-1}gh = e$.

Notice the reversal in order of the elements. This result can be extended to give

$$(g_1 g_2 \dots g_k)^{-1} = g_k^{-1} g_{k-1}^{-1} \dots g_1^{-1},$$

for any $k$ elements $g_1, g_2, \dots, g_k$ of a group.

## Cancellation laws

Suppose that $G$ is a group and that $g, h, k \in G$ with $gh = gk$. Then we can cancel $g$ and obtain $h = k$. To see this, premultiply both sides of the equation $gh = gk$ by $g^{-1}$ to obtain $g^{-1}gh = g^{-1}gk$. Now $g^{-1}gh = eh = h$ and $g^{-1}gk = ek = k$ and so $h = k$. Similarly, if $hg = kg$ then $h = k$.

The laws

$$gh = gk \Rightarrow h = k$$
$$\text{and } hg = kg \Rightarrow h = k$$

are called the **cancellation laws**.

## Latin square property

If $G$ is a finite group, its binary operation can be displayed in a Cayley table as for $D_4$ in Chapter 1.

| $G$ | $h$ |
|---|---|
| $g$ | $gh$ |

| $G$ | $a$ | $a^{-1}b$ |
|---|---|---|
| $a$ | | $b$ |
| $ba^{-1}$ | $b$ | |

Let $G$ be a group and let $a, b \in G$. Then there is an element $x \in G$ such that $ax = b$, namely $x = a^{-1}b$. By the cancellation law, $x$ is the only such element of $G$. Similarly, there is a unique element $y$ of $G$, namely $y = ba^{-1}$, such that $ya = b$. In terms of the Cayley table, this says that $b$ appears exactly once in the row for $a$ and exactly once in the column for $a$. This property of a square array, that each element appears exactly once in each row and exactly once in each column is called the **Latin square** property.

### EXERCISE 2

Let $G$ be a group and let $g \in G$. Show that if $g^2 = g$ then $g = e$.

### EXERCISE 3

In each of the following you are given two elements $a, b$ of a group $G$. In each case, find $x$ and $y$ in $G$ such that $ax = b = ya$.

(i) $G$ is the orthogonal group $O_2$; $a = \text{rot}_{\frac{\pi}{6}}$, $b = \text{ref}_{\frac{\pi}{2}}$.

(ii) $G$ is the dihedral group $D_4$; $a = r_1$ (rotation through $\frac{\pi}{2}$), $b = s_1$ (reflection in the $x$-axis).

(iii) $G$ is the symmetric group $S_4$; $a = (1\ 2\ 4)$, $b = (1\ 3)$.

### EXERCISE 4

In $D_4$, find $g, h, k$ with $gh = kg$ and $h \neq k$.

## EXERCISE 5

Let $G$ be a group in which every element is its own inverse, that is $a^{-1} = a$ for all $a \in G$. Show that $G$ is abelian. (Hint: $ab = (ab)^{-1} = \ldots$)

# 4.6 Direct Products

In this section we look at a way of building a new group from two, possibly equal, given groups. Before we do this in general, we look at an example. The Euclidean plane $\mathbb{R}^2$ is a group $(\mathbb{R}^2, +)$ under componentwise addition

$$(a, b) + (c, d) = (a + c, b + d).$$

The neutral element is clearly $(0, 0)$ and the inverse of $(a, b)$ is $(-a, -b)$.

This is a special case (G=H=$\mathbb{R}$) of the following construction of a new group from two given groups. Let $G$ and $H$ be groups. The **Cartesian product** $G \times H$ consists of all ordered pairs $(g, h)$, where $g \in G$ and $h \in H$. We define a binary operation on $G \times H$ componentwise as follows

$$(g_1, h_1)(g_2, h_2) = (g_1 g_2, h_1 h_2).$$

For example, in $D_4 \times D_4$, $(s_1, r_2)(s_2, r_1) = (r_3, r_3)$.

We check each of the group axioms for $G \times H$.

**G1:** $G \times H$ is closed because $G$ and $H$ are both closed.

**G2:** Let $g_1, g_2, g_3 \in G$ and $h_1, h_2, h_3 \in H$. Then, since $G$ and $H$ both satisfy the associative law,

$$\begin{aligned} ((g_1, h_1)(g_2, h_2))(g_3, h_3) &= (g_1 g_2, h_1 h_2)(g_3, h_3) \\ &= ((g_1 g_2) g_3, (h_1 h_2) h_3) \\ &= (g_1 (g_2 g_3), h_1 (h_2 h_3)) \\ &= (g_1, h_1)(g_2 g_3, h_2 h_3) \\ &= (g_1, h_1)((g_2, h_2)(g_3, h_3)). \end{aligned}$$

This shows that $G \times H$ satisfies the associative law.

**G3:** Let $g \in G$ and $h \in H$. Then

$$(g, h)(e_G, e_H) = (g e_G, h e_H) = (g, h) = (e_G g, e_H h) = (e_G, e_H)(g, h).$$

This shows that $(e_G, e_H)$ is a neutral element for $G \times H$.

**G4:** Let $g \in G$ and $h \in H$. Then

$$(g, h)(g^{-1}, h^{-1}) = (g g^{-1}, h h^{-1}) = (e_G, e_H) = (g^{-1} g, h^{-1} h) = (g^{-1}, h^{-1})(g, h).$$

This shows that $(g, h)$ has inverse $(g^{-1}, h^{-1})$ in $G \times H$.

The group $(G \times H)$ is called the **direct product** of $G$ and $H$.

When $G$ and $H$ are in additive notation, the definition of the binary operation in $G \times H$ becomes

$$(g_1, h_1) + (g_2, h_2) = (g_1 + g_2, h_1 + h_2).$$

If either $G$ or $H$ is infinite, $|G \times H| = \infty$ and if both are finite then $|G \times H| = |G||H|$, there being $|G|$ choices for the component $g$ in $(g, h)$ and, for each of these, $|H|$ choices for $h$.

**EXERCISE 6**

Show that if $G$, $H$ are both abelian then so is $G \times H$.

## 4.7 Further Exercises on Chapter 4

**EXERCISE 7**

Show that $(\mathbb{R}\backslash\{-1\}, \dagger)$ is a group where, for $a, b \in \mathbb{R}$, $a \dagger b = ab + a + b$.
Explain why $(\mathbb{R}, \dagger)$ is not a group.

**EXERCISE 8**

Let $G$ be a group and let $g, h, k \in G$.

(i) Show that if $ghg = gkg$ then $h = k$.

(ii) Give an example to show that it is not always true that if $hgh = kgk$ then $h = k$.

**EXERCISE 9**

Let $G$ denote the set of all ordered pairs $(a, b)$ where $a$ and $b$ are real numbers and $b$ is non-zero. For $(a_1, b_1), (a_2, b_2) \in G$, set

$$(a_1, b_1) \oplus (a_2, b_2) = (a_1 + b_1 a_2, b_1 b_2).$$

(i) Compute $(2, -1) \oplus (3, 1)$, $(2, -1) \oplus (2, -1)$ and $(1, 2) \oplus (\frac{-1}{4}, \frac{1}{2})$.

(ii) Show that $G$ is a group under $\oplus$. Is $G$ abelian?

**EXERCISE 10**

Let $G$ denote the set of all ordered pairs $(a, b)$ where $a$ is a non-zero real number and $b = \pm 1$. For $(a_1, b_1), (a_2, b_2) \in G$, set

$$(a_1, b_1) \odot (a_2, b_2) = (a_1 a_2^{b_1}, b_1 b_2).$$

(i) Compute $(2, -1) \odot (3, 1)$, $(2, -1) \odot (2, -1)$ and $(2, 1) \odot (\frac{1}{2}, 1)$.

(ii) Show that $G$ is a group under $\odot$.

(iii) Is $G$ abelian?

(iv) Which elements of $G$ are their own inverses?

## EXERCISE 11

Let $G$ be the group $\{1, -1\}$ under multiplication. Complete the given Cayley table for the direct product $G \times G$.

| $G \times G$ | $(1, 1)$ | $(1, -1)$ | $(-1, 1)$ | $(-1, -1)$ |
|---|---|---|---|---|
| $(1, 1)$ | | | | |
| $(1, -1)$ | | | | |
| $(-1, 1)$ | | | | |
| $(-1, -1)$ | | | | |

# 5 • Subgroups

In several of the examples we have met, one group is to be been found inside another. For instance, the group of symmetries of the rectangle is found inside $D_4$, the group of symmetries of a square, which, in turn, is inside $O_2$, the orthogonal group or group of symmetries of the circle. Similarly, the group $(\mathbb{R}\backslash\{0\}, \times)$, the non-zero real numbers under multiplication, is inside $(\mathbb{C}\backslash\{0\}, \times)$. In each of these cases the binary operation, composition for the groups of symmetries and multiplication in the groups of numbers, is the same in the smaller group as in the bigger one. These are examples of subgroups. The notion of subgroups, is an important aid in our understanding of groups.

## 5.1 Subgroups

Let $G$ be a group. We say that a subset $H$ of $G$ is a **subgroup** of $G$ if $H$ is a group under the same binary operation as $G$.

A very easy example of a subgroup of any group $G$ is the group consisting of just the neutral element $\{e\}$. This is known as the **trivial subgroup**. It is also clear that any group $G$ is a subgroup of itself. Quite often these two subgroups need to be excluded from consideration. The term **proper subgroup** is used to describe any subgroup which is not the whole group and the term **non-trivial subgroup** is used to describe any subgroup other than $\{e\}$.

### Neutral elements and inverses in subgroups

If $H$ is a subgroup of $G$ then each has a neutral element, $e_G$ in $G$ and $e_H$ in $H$, and each element $h$ of $H$ has an inverse in $H$ and an inverse in $G$. Before we proceed further, we check that $e_H = e_G$ and that the inverses of $h$ in $G$ and $H$ are equal. To see that $e_H = e_G$, observe that $e_H e_H = e_H$ and that, because $e_H \in G$ where $e_G$ is neutral, $e_G e_H = e_H$. Thus, $e_H e_H = e_G e_H$ and, by cancellation of $e_H$ in $G$, it follows that $e_H = e_G$.

If $a$ is the inverse of $h$ in $H$ then $ah = ha = e_H = e_G$ so $a$ must be the inverse of $h$ in $G$ as well as in $H$. Thus, when considering subgroups, no ambiguity arises in the use of the notation $h^{-1}$.

### The subgroup criterion

Before we look at some more examples of subgroups, we shall prove a theorem which will make it easier to check whether a given subset is a subgroup or not. At the moment we would need to check all the group axioms. This is unnecessary as the following theorem shows. The essential difference in checking the criterion in the theorem and checking the group axioms lies in the fact that the associative law G2 automatically holds in any subset of a group and need not be checked.

• *Theorem 1 (The subgroup criterion)*

Let $G$ be a group and $H$ be a subset of $G$. $H$ is a subgroup of $G$ if and only if the following three conditions are satisfied.

**SG1:** $H$ is non-empty.

**SG2:** $H$ is closed under the binary operation for $G$. In other words if $g \in H$ and $h \in H$ then $gh \in H$.

**SG3:** The inverse of each element of $H$ belongs to $H$. In other words if $h \in H$ then $h^{-1} \in H$. ●

PROOF

We need to prove two things. Firstly, we must show that if $H$ satisfies SG1, SG2 and SG3 then $H$ is a subgroup. Suppose then that $H$ satisfies SG1, SG2 and SG3. We have to check the group axioms to show that $H$ is a group. SG2 shows that $H$ is closed under the binary operation on $G$ so G1 holds. Since the associative law $(ab)c = a(bc)$ holds for all $a, b, c \in G$ and $H \subseteq G$, it holds for all $a, b, c \in H$. Thus, $H$ satisfies G2. By SG1, there is at least one element $h \in H$. By SG3, $h^{-1} \in H$ and, by SG2, $e_G = hh^{-1} \in H$. Since $e_G$ is neutral in $G$ and is in $H$, it is neutral in $H$. Thus, G3 holds for $H$ with neutral element $e_G$. Finally, let $h \in H$. Then $h^{-1} \in H$ by SG3 and is an inverse for $h$ in $H$. Thus, $H$ satisfies G4. Since $H$ is a subset of $G$ and satisfies the group axioms for the binary operation on $G$ it is a subgroup of $G$.

We now need to show that if $H$ is a subgroup then it satisfies SG1, SG2 and SG3. Suppose then that $H$ is a subgroup of $G$; that is, $H$ is a subset of $G$ and is a group under the binary operation in $G$. It is part of the definition of a group that it is non-empty so SG1 is satisfied. Since $H$ is a group, it is closed under the binary operation on $G$ so SG2 is satisfied by $H$. Let $h \in H$. We have seen that $h^{-1}$ is the inverse of $h$ in $H$ as well as in $G$ so SG3 holds. ●

## 5.2 Examples of Subgroups

Several of the examples listed below are important as groups in their own right not just as subgroups. It is, however, easier to prove that they are subgroups of a given group rather than directly check the group axioms. In the first example, we include some subsets which are not subgroups.

### Positive reals

Consider the following subsets of the group $(\mathbb{R}\backslash\{0\}, \times)$:
$H_1 = \{x \in \mathbb{R} : x > 0\}$ (the positive reals);
$H_2 = \{x \in \mathbb{R} : x > 1\}$;
$H_3 = \{x \in \mathbb{R} : x < 0\}$ (the negative reals).
$H_1$ satisfies SG1,2,3 and is a subgroup of $\mathbb{R}\backslash\{0\}$. We shall denote this subgroup by $\mathbb{R}^+$.

$H_2$ satisfies SG1,2 but SG3 fails as $2 \in H_2$ and $2^{-1} \notin H_2$. $H_3$ fails SG2: $-1, -2 \in H_3$ but $(-1)(-2) = 2 \notin H_3$. Thus, $H_2$ and $H_3$ are not subgroups of $(\mathbb{R}\backslash\{0\}, \times)$.

### Powers of 2

We show that the set $H = \{2^n : n \in \mathbb{Z}\}$ of all powers of 2 is a subgroup of $(\mathbb{R}\backslash\{0\}, \times)$.

**SG1**: $H$ is non-empty because $1 = 2^0 \in H$.

**SG2**: Let $g, h \in H$. Then $g = 2^m$ and $h = 2^n$ for some $m, n \in \mathbb{Z}$ and $gh = 2^m 2^n = 2^{m+n} \in H$.

**SG3**: Let $h \in H$. Then $h = 2^n$ for some $n \in \mathbb{Z}$ and $h^{-1} = 2^{-n} \in H$.

Thus, by the subgroup criterion, $H$ is a subgroup of $(\mathbb{R}\backslash\{0\}, \times)$. This subgroup, built from the single element 2, is an example of what we shall later call a **cyclic subgroup**.

## Roots of unity

Fix a positive integer $n$ and, in $\mathbb{C}\backslash\{0\}$ under multiplication, let $H = \{z \in \mathbb{C} : z^n = 1\}$. Thus, $H$ is the set of all complex $n$th roots of unity. For example, if $n = 4$ then $H = \{1, i, -1, -i\}$. We check SG1,SG2,SG3 as follows.

**SG1**: $H$ is non-empty because $1 \in H$.

**SG2**: Let $a, b \in H$, that is $a^n = b^n = 1$. Then $(ab)^n = a^n b^n = 1$ so $ab \in H$.

**SG3**: Let $a \in H$. Then $(a^{-1})^n = (a^n)^{-1} = 1^{-1} = 1$ so $a^{-1} \in H$.

By the subgroup criterion, $H$ is a subgroup of $\mathbb{C}\backslash\{0\}$ under multiplication. It is called the **group of $n$th roots of unity** and we shall write it as $U_n$.

$U_1 = \{1\}$, $U_2 = \{1, -1\}$, $U_3 = \{1, \mathrm{e}^{\frac{2\pi \mathrm{i}}{3}}, \mathrm{e}^{\frac{4\pi \mathrm{i}}{3}}\}$, $U_4 = \{1, \mathrm{i}, -1, -\mathrm{i}\}, \ldots$. In general, $U_n = \{1, \mathrm{e}^{\frac{2\pi \mathrm{i}}{n}}, \mathrm{e}^{\frac{4\pi \mathrm{i}}{n}}, \ldots, \mathrm{e}^{\frac{2(n-1)\pi \mathrm{i}}{n}}\}$ and the order of $U_n$ is $n$.

## Multiples of $m$

For a group in additive notation, SG2 and SG3 become

**SG2**: $a + b \in H$ for all $a, b \in H$ and **SG3**:$-a \in H$ for all $a \in H$.

Fix a positive integer $m$ and, in $\mathbb{Z}$ under addition, let $m\mathbb{Z}$ denote the set of all multiples of $m$, that is $m\mathbb{Z} = \{a \in \mathbb{Z} : \text{ there exists } n \in \mathbb{Z} \text{ with } a = mn\}$, e.g. $5\mathbb{Z} = \{\ldots, -15, -10, -5, 0, 5, 10, 15, \ldots\}$.

**SG1**: $0 = m \times 0 \in m\mathbb{Z}$ so $m\mathbb{Z} \neq \emptyset$.

**SG2**: Let $a, b \in n\mathbb{Z}$; $a = mn, b = mk$ for some $n, k \in \mathbb{Z}$. Then $a + b = m(n + k) \in m\mathbb{Z}$.

**SG3**: Let $a = mn \in m\mathbb{Z}$. Then $-a = m(-n) \in m\mathbb{Z}$.

By the subgroup criterion, $m\mathbb{Z}$ is a subgroup of $\mathbb{Z}$.

## The special linear group

Let $n \geqslant 1$. In the general linear group $GL_n(\mathbb{R})$, let $H$ denote the set of all $n \times n$ real matrices with determinant 1, that is, $H = \{A \in GL_n(\mathbb{R}) : \det A = 1\}$. Note that, by Theorem 1(i) of Chapter 3, $H$ is a sub*set* of $GL_n(\mathbb{R})$.

**SG1**: $H \neq \emptyset$ because $I_n \in H$.

**SG2**: Let $A, B \in H$, that is $\det A = \det B = 1$. Then $\det(AB) = \det A \det B = 1$. Hence, $AB \in H$.

**SG3**: Let $A \in H$. Then $\det A \det A^{-1} = \det(AA^{-1}) = \det I_n = 1$ so $\det A^{-1} = 1/\det A = 1/1 = 1$. Hence, $A^{-1} \in H$.

By the subgroup criterion, $H$ is a subgroup of $GL_n(\mathbb{R})$. It is called the **special linear group** and is written $SL_n(\mathbb{R})$. There are similarly defined special linear groups $SL_n(\mathbb{Q})$ and $SL_n(\mathbb{C})$ over $\mathbb{Q}$ and $\mathbb{C}$.

## The special orthogonal group

We show that the set $H$ of all rotations in the plane about the origin, that is $H = \{\mathrm{rot}_\theta : \theta \in \mathbb{R}\}$, is a subgroup of $O_2$.

**SG1**: Clearly $H$ is non-empty since $\text{rot}_0 \in H$.

**SG2**: From Table 1.1 we know that $\text{rot}_\alpha \, \text{rot}_\beta = \text{rot}_{\alpha+\beta}$.

**SG3**: This follows since $\text{rot}_{-\alpha}$ is the inverse of $\text{rot}_\alpha$.

This group is denoted by $SO_2$ and is called the **special orthogonal group**.

## Orthogonal matrices

Recall from Chapter 3 that a $2 \times 2$ real matrix $A$ is said to be orthogonal if $A^T = A^{-1}$. Let $H$ denote the set of all such matrices and note that this is a subset of $GL_2(\mathbb{R})$. We show below that $H$ is a subgroup of $GL_2(\mathbb{R})$.

**SG1**: Clearly $H$ is non-empty since $I_2 \in H$.

**SG2**: Let $A, B \in H$, that is $A^T = A^{-1}$ and $B^T = B^{-1}$. Now $(AB)^T = B^T A^T = B^{-1}A^{-1} = (AB)^{-1}$. Thus, $AB \in H$.

**SG3**: This follows since if $A^T = A^{-1}$ then $(A^{-1})^T = (A^T)^T = A = (A^{-1})^{-1}$.

This group is, in a sense to be explained in Chapter 9, the same as the group of symmetries of the circle to which we have already given the name **orthogonal group**.

### EXERCISE 1

In each of the following you are given a group $G$ and a subset $H$ of $G$. In each case, decide whether $H$ is a subgroup of $G$.

(i) $G = S_4$; $H = \{f \in S_4 : f(4) = 4\}$, the set of those permutations in $S_4$ which send 4 to 4.

(ii) $G = S_4$; $H = \{f \in S_4 : f(4) = 4 \text{ or } 3\}$, the set of those permutations in $S_4$ which send 4 either to 4 or 3.

(iii) $G = (\mathbb{R}, +)$; $H = \{x \in \mathbb{R} : \sin x = 0\}$.

(iv) $G = (\mathbb{R}, +)$; $H = \{x \in \mathbb{R} : \cos x = 0\}$.

### EXERCISE 2

Let $H = \left\{ \begin{pmatrix} 1 & a \\ 0 & 1 \end{pmatrix} : a \in \mathbb{R} \right\}$, the set of $2 \times 2$ matrices representing shears parallel to the $x$-axis. Show that $H$ is a subgroup of $GL_2(\mathbb{R})$. Is $H$ abelian?

## 5.3 Groups of Symmetries

In Chapter 1 we introduced various groups of symmetries. Here we give a general description of such groups as subgroups of the orthogonal group $O_2$; that is, the group of all rotations $\text{rot}_\theta$ and all reflections $\text{ref}_\theta$.

Let $A$ be a figure (such as a circle or square) in $\mathbb{R}^2$, positioned in such a way that its centre lies at the origin. Let $f \in O_2$ be a rotation or reflection. Then the **image**, $f(A)$, of $A$ under $f$ is the set $\{f(a) : a \in A\}$ of all points that are obtained by applying $f$ to points of $A$. Figure 5.1 shows the images of a rectangle for $f_1 = \text{rot}_\pi$ and for $f_2 = \text{rot}_{\frac{\pi}{2}}$.

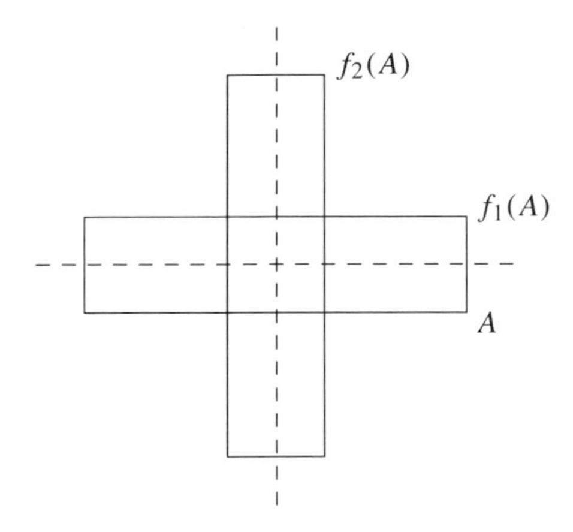


**Fig 5.1** Images of a rectangle.

Notice that in Fig 5.1 both images are rectangles but one occupies the same position as the original whereas the other does not. In other words, $f_1(A) = A$ but $f_2(A) \neq A$. We say that $f$ is a **symmetry** of $A$ if $f(A) = A$. Thus, a symmetry of $A$, although it may move the individual points of $A$, leaves $A$ in its original position.

We have already looked at the symmetries of rectangles, squares and circles.

- The symmetries of the non-square rectangle are $\text{rot}_0$, $\text{rot}_\pi$, $\text{ref}_0$ and $\text{ref}\,\pi$.
- The symmetries of the square are the above four symmetries together with $\text{rot}_{\frac{\pi}{2}}$, $\text{rot}_{\frac{3\pi}{2}}$, $\text{ref}_{\frac{\pi}{2}}$ and $\text{ref}_{\frac{3\pi}{2}}$.
- Every element of the orthogonal group $O_2$ is a symmetry of the circle.

## Theorem 2

Let $A$ be a figure with centre at the origin. Let $G$ be the set of all symmetries of $A$. Then $G$ is a subgroup of the orthogonal group $O_2$.

PROOF

We apply the subgroup criterion.

**SG1:** $G$ is non-empty since $\text{rot}_0$ is a symmetry of $A$.

**SG2:** Let $f, g$ be symmetries of $A$. Thus, $f(A) = A$ and $g(A) = A$. Now $fg(A) = f(g(A)) = f(A) = A$ and so $fg$ is a symmetry of $A$.

**SG3:** Let $f$ be a symmetry of $A$. Thus, $f(A) = A$. Then $f^{-1}(A) = f^{-1}(f(A)) = (f^{-1}f)(A) = \text{id}(A) = A$ and so $f^{-1}$ is a symmetry of $A$.

By the subgroup criterion, $G$ is a subgroup of $O_2$.

The group of all symmetries of a figure $A$ is called the **group of symmetries** of $A$.

## The dihedral groups

Here we consider an $n$-sided regular polygon $P_n$. Such a polygon has $n$ rotations which are symmetries. The nature of the reflections of $P_n$ depends on whether $n$ is odd or even. This is illustrated in Fig 5.2.

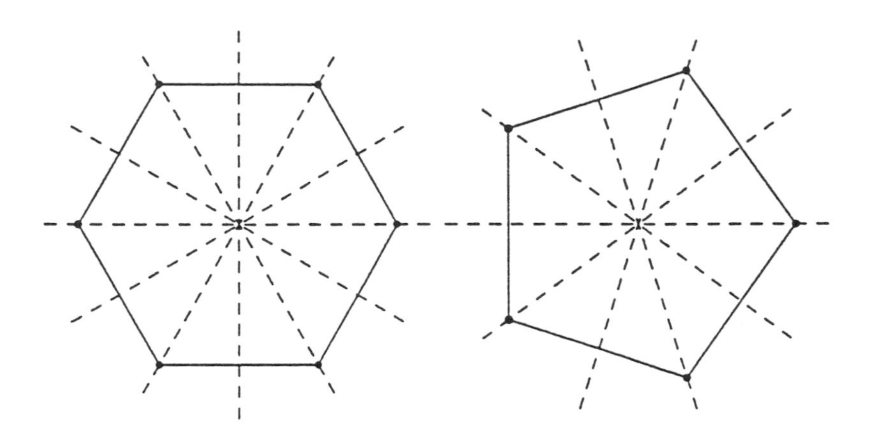

**Fig 5.2** Symmetries of regular polygons.

When $n$ is even, $P_n$ has two types of lines of reflection, those through opposite vertices, of which there are $\frac{n}{2}$, and those through midpoints of opposite edges, of which there are again $\frac{n}{2}$. However, if $n$ is odd, all the lines of reflection for $P_n$ go through a vertex and the midpoint of the opposite edge. There are $n$ such lines. In both the even and odd cases there are $n$ reflections. If $P_n$ is placed so that the $x$-axis is a line of reflection then the group of symmetries of $P_n$ is $\{\text{rot}_{\frac{2k\pi}{n}}, \text{ref}_{\frac{2k\pi}{n}} : k = 0 \ldots n-1\}$ and has order $2n$. This group is called the **dihedral group** $D_n$. In particular, $D_4$ denotes the group of symmetries of the square.

## Rotation groups

Observe that the group of symmetries of each of the shapes in Fig 5.3 consists only of

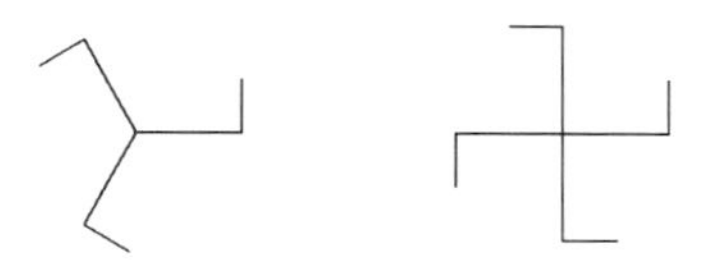

**Fig 5.3** Shapes with only rotational symmetry.

rotations. For a given geometrical figure, $A$, consider the subset of the group, $H$, of symmetries of $A$ consisting of all the rotations of $A$. This is the intersection of two subgroups of the group $O_2$, namely $H$ and the special orthogonal group $SO_2$. By Theorem 3 below, it forms a subgroup of $O_2$. This subgroup is called the **rotation group** of the figure. For example, the rotation group of the square is $\{e, \text{rot}_{\frac{\pi}{2}}, \text{rot}_{\pi}, \text{rot}_{\frac{3\pi}{2}}\}$ and the rotation group of the circle is the special orthogonal group $SO_2$. For the shapes in Fig 5.3, the rotation group is equal to the group of symmetries.

● *Theorem 3*

Let $H$ and $K$ be two subgroups of a group $G$. Then $H \cap K$ is a subgroup of $G$. ●

PROOF

Both $H$ and $K$ have the same neutral element as $G$. Thus, $e_G \in H \cap K$ so $H \cap K \neq \emptyset$ and SG1 holds for $H \cap K$. Now let $a, b \in H \cap K$. Both $H$ and $K$ satisfy the subgroup criterion so $ab \in H, a^{-1} \in H, ab \in K$ and $a^{-1} \in K$. Therefore, $ab \in H \cap K$ and $a^{-1} \in H \cap K$. Thus, in addition to SG1, $H \cap K$ satisfies SG2 and SG3. By the subgroup criterion, $H \cap K$ is a subgroup of $G$. ●

**EXERCISE 3**

Consider the 26 capital letters of the alphabet, each written as symmetrically as possible. For example

Which letters have

(i) group of symmetries of order 1 (the only symmetry is the identity);

(ii) group of symmetries of order 2 consisting of the identity and a reflection;

(iii) group of symmetries of order 2 consisting of the identity and a rotation;

(iv) group of symmetries of order 4;

(v) group of symmetries equal to $D_3$;

(vi) group of symmetries equal to $D_4$;

(vii) group of symmetries equal to $O_2$?

## 5.4 Further Exercises on Chapter 5

**EXERCISE 4**

Show that $H = \left\{\begin{pmatrix} a & b \\ -b & a \end{pmatrix} : a, b \in \mathbb{R} \text{ and } ab \neq 0\right\}$ is a subgroup of $GL_2(\mathbb{R})$. Is $H$ abelian?

**EXERCISE 5**

Show that $H = \left\{\begin{pmatrix} a & b \\ b & a \end{pmatrix} : \quad a, b \in \mathbb{R} \text{ and } a^2 - b^2 = 1\right\}$ is a subgroup of $GL_2(\mathbb{R})$. Is $H$ abelian?

## EXERCISE 6

Let $t$ be a non-zero element of $\mathbb{R}$. Show that
$H = \left\{ \begin{pmatrix} \cos\theta & -t\sin\theta \\ \frac{1}{t}\sin\theta & \cos\theta \end{pmatrix} : \theta \in \mathbb{R}, \right\}$ is a subgroup of $GL_2(\mathbb{R})$. Is $H$ abelian?

## EXERCISE 7

Sketch a figure which has a group of symmetries of order 5.

## EXERCISE 8

Let $G \times H$ be the direct product of two groups $G$ and $H$. In a sense, $G \times H$ contains copies of each of the groups $G$ and $H$. The 'copy' of $G$ consists of those pairs $(g, h)$ in which the element $h$ of $H$ is $e_H$, that is $G' = \{(g, e_H) : g \in G\}$. Similarly, $H' = \{(e_G, h) : H\}$ can be thought of as a copy of $H$ in $G \times H$.

Show that $G'$ and $H'$ are subgroups of $G \times H$.

## EXERCISE 9

Show that, for any group $G$, the set $D = \{(g, g) : g \in G\}$ is a subgroup of $G \times G$. (This subgroup $D$ is called the **diagonal** subgroup of $G \times G$.) When $G = (\mathbb{R}, +)$, describe the subgroup $D$ geometrically.

## EXERCISE 10

Let $G$ be an abelian group. Show that the set $\{g \in G : g^2 = e\}$ is a subgroup of $G$.

## EXERCISE 11

Let $G$ be the orthogonal group $O_2$. Show that the set $\{g \in G : g^2 = e\}$ is not a subgroup of $G$.

## EXERCISE 12

Let $G$ be the symmetric group $S_3$. Show that the set $\{g \in G : g^2 = \mathrm{id}\}$ is not a subgroup of $G$. Is the set $\{g \in G : g^3 = \mathrm{id}\}$ a subgroup? What happens in $S_4$?

## EXERCISE 13

Investigate the subgroups of $S_3$, $D_4$, and $S_4$. How many can you find in each?

# 6 • Cyclic Groups

A cyclic group is a group which can be built from a single element by taking powers. Any element of any group can be used to build a cyclic subgroup so cyclic groups appear inside all groups. This chapter establishes the basic properties of these important groups.

## 6.1 Cyclic Groups

Consider the rotation group $G = \{e, \text{rot}_{\frac{2\pi}{5}}, \text{rot}_{\frac{4\pi}{5}}, \text{rot}_{\frac{6\pi}{5}}, \text{rot}_{\frac{8\pi}{5}}\}$ of a regular pentagon. The following features of this group are typical of the groups to be studied in this chapter.

- The elements of $G$ are all powers, $g^0, g^1, g^2, g^3$ and $g^4$ of the element $g = \text{rot}_{\frac{2\pi}{5}}$.
- They are the only powers of $g$. Continuing to take powers does not give any new rotations, $g^6 = g$, $g^7 = g^2$ etc. In general, $g^m = g^r$ where $r$ is the remainder on division of $m$ by 5. For example, $g^{45} = g^0 = e$, $g^{77} = g^2$ and $g^{-2} = g^3$.
- The order of $G$, 5, is the least positive integer $n$ such that $g^n = e$.

Let $G$ be a group. If there is an element $g \in G$ such that every element of $G$ has the form $g^n$ for some $n \in Z$ then we say that $G$ is a **cyclic group**. Any such element $g$ is called a **generator** of $G$.

### Example 1

The rotation group $G$ above is cyclic with generator $g = \text{rot}_{\frac{2\pi}{5}}$. Notice that $g^2$ is also a generator for $G$ because $(g^2)^2 = g^4$, $(g^2)^3 = g^6 = g$, $(g^2)^4 = g^8 = g^3$ and $(g^2)^5 = g^{10} = e$ so that all five elements of $G$ are powers of $g^2$. As an exercise, you should check that $g^3$ and $g^4$ are also generators of $G$.

### Example 2

For each positive integer $n$, the rotation group of the regular $n$-gon is cyclic with generator $\text{rot}_{\frac{2\pi}{n}}$.

### Example 3

For each positive integer $n$, the group $U_n$ of $n$th roots of unity is cyclic with generator $e^{\frac{2\pi i}{n}}$. This example appears to be closely related to the previous one. We will show how to make this idea precise when we consider isomorphism in Chapter 9.

### Example 4

A group $G$ in additive notation is cyclic when it has an element $g$ such that every element has the form $ng$ for some $n \in \mathbb{Z}$. For example, if we take $g = 1$ in $\mathbb{Z}$, every element $n$ of $\mathbb{Z}$ can be written as $ng$. Thus, the group $(\mathbb{Z}, +)$ is cyclic with generator 1.

● *Theorem 1*

Every cyclic group $G$ is abelian. ●

PROOF

Let $g$ be a generator of $G$ and let $a, b \in G$. There exist integers $m, n$ such that $a = g^m$ and $b = g^n$. Then

$$ab = g^m g^n = g^{m+n} = g^{n+m} = g^n g^m = ba.$$

Thus, $ab = ba$ for all $a, b \in G$ and so $G$ is abelian. ●

The converse to the above theorem is false; there are abelian groups which are not cyclic. For example, let $K = \{e, r, s, t\}$ be the group of symmetries of the non-square rectangle. It is clear from the Cayley table in Fig 4.1 that $K$ is abelian. The only power of $e$ is $e$ itself and, if $g$ is any one of the other three elements, the only powers of $g$ are $e$ and $g$. Thus, $K$ is not cyclic.

**EXERCISE 1**

(i) Let $G$ be the rotation group of a regular hexagon. Which elements of $G$ are generators of $G$?

(ii) Which elements of $(\mathbb{Z}, +)$ are generators of $(\mathbb{Z}, +)$?

## 6.2 Cyclic Subgroups

Inside any group $G$, we can use any element $g$ to generate a cyclic subgroup. We use the subgroup criterion to check that the set $H = \{g^n : n \in \mathbb{Z}\}$ of all powers of $g$ is a subgroup of $G$.

**SG1:** Since $g \in H$, $H$ is non-empty.

**SG2:** Let $h, k \in H$. There are integers $m, n$ with $h = g^m$ and $k = g^n$. Therefore, $hk = g^m g^n = g^{m+n} \in H$.

**SG3:** Let $h = g^m \in H$. Then $h^{-1} = g^{-m} \in H$.

By the subgroup criterion, $H$ is a subgroup of $G$. We call this subgroup the **cyclic subgroup generated by** $g$. It will be denoted by $< g >$. It is the smallest subgroup of $G$ containing $g$ in the sense that if $K$ is any subgroup of $G$ with $g \in K$ then $< g > \subseteq K$.

### Examples of cyclic subgroups

(i) The cyclic subgroup of $D_n$ generated by the element $\text{rot}_{\frac{2\pi}{n}}$ is the rotation group of the regular $n$-sided polygon and has order $n$.

(ii) Any reflection $s$ in $D_n$ generates a cyclic subgroup of $D_n$. This subgroup has just two elements since $s^2 = e, s^3 = s, s^4 = e, \ldots$ and $s^{-1} = s, s^{-2} = e, \ldots$. Thus, $< s > = \{e, s\}$.

(iii) The cyclic subgroup of $(\mathbb{R}\backslash\{0\}, \times)$ generated by 2 is
$< 2 > = \{2^n : n \in \mathbb{Z}\} = \{\ldots, \frac{1}{8}, \frac{1}{4}, \frac{1}{2}, 1, 2, 4, 8, \ldots\}$.

(iv) In $(\mathbb{C}\backslash\{0\}, \times)$ the cyclic subgroup generated by the element $e^{\frac{2\pi i}{n}}$ is the group $U_n$ of $n$th roots of unity, e.g. $<i>=\{1, i, -1, -i\}$.

(v) For a group in additive notation, $< g >$ consists of all elements of the form $ng$, $n \in \mathbb{Z}$. For example, the cyclic subgroup $< 2 >$ of $(\mathbb{Z}, +)$ is $2\mathbb{Z} = \{\ldots, -4, -2, 0, 2, 4, \ldots\}$; that is, the even integers or multiples of 2. For each $m \in \mathbb{Z}$, the cyclic subgroup of $(\mathbb{Z}, +)$ generated by $m$ is $m\mathbb{Z} = \{\ldots, -2m, -m, 0, m, 2m, \ldots\}$. In particular, $< 1 >=< -1 >$ is the whole group $\mathbb{Z}$.

**EXERCISE 2**

In each of the following, decide whether $< g >$ is finite. If so, list its elements and, if not, give a formula for a typical element $g^n$.

(i) (1 2 3 4) in $S_4$.

(ii) $\begin{pmatrix} 0 & 1 \\ 1 & 0 \end{pmatrix}$ in $GL_2(\mathbb{R})$.

(iii) $\begin{pmatrix} 1 & 1 \\ 0 & 1 \end{pmatrix}$ in $GL_2(\mathbb{R})$.

## 6.3 Order of Elements

Some of the cyclic groups we have seen, such as the rotation group of a regular polygon, are finite, while others, such as $\mathbb{Z}$ and the powers of 2, are infinite. How do we determine $| < g > |$? For the case of the rotation group $G =< g >$ of a regular pentagon, where $g = \mathrm{rot}_{\frac{2\pi}{5}}$, $|G| = 5$ which is the least positive integer $n$ such that $g^n = e$. On the other hand, in $(\mathbb{R}\backslash\{0\}, \times)$, there is no positive integer $n$ with $2^n = 1$ and $| < 2 > | = \infty$.

For an element $g$ of a group $G$, we define the **order** of $g$ to be the least positive integer $n$ such that $g^n = e$, if such integers exist, and to be infinite if no such integers exist.

Thus, the order of $\mathrm{rot}_{\frac{2\pi}{5}}$ in the group $D_5$ is 5 and the order of 2 in $(\mathbb{R}\backslash\{0\}, \times)$ is $\infty$. In the symmetric group $S_n$ a cycle of length $k$ has order $k$. For example, (1 2) and (1 2 3) have orders 2 and 3 respectively.

**EXERCISE 3**

Find the order of each the following elements of $D_6$: $\mathrm{rot}_{\frac{\pi}{3}}$, $\mathrm{rot}_{\frac{2\pi}{3}}$, $\mathrm{rot}_{\pi}$, $\mathrm{ref}_{\pi}$ .

Computing the order of a cyclic subgroup $< g >$ requires knowledge of when two powers $g^i$ and $g^j$ of $g$ are equal. Now

$$g^i = g^j \Leftrightarrow g^i g^{-j} = e \Leftrightarrow g^{i-j} = e.$$

If $g$ has infinite order and $i > j$ then $g^{i-j} \neq e$, because $i - j > 0$, and so $g^i \neq g^j$. Thus, in this case, the elements $g^i$, $i \in \mathbb{Z}$ are all distinct and we have the following theorem.

## • Theorem 2

Let $G$ be a group and let $g \in G$ be an element of infinite order. Then the cyclic subgroup $< g >$ of $G$ generated by $g$ has infinite order. ●

The situation for an element of finite order is different. For example, in $D_5$, the element $\text{rot}_{\frac{2\pi}{5}}$ has exactly five different powers.

## • Theorem 3

Let $G$ be a group and let $g \in G$ be an element of finite order $n$.

(i) For $m \in \mathbb{Z}$, $g^m = g^r$ where $r$ is the remainder on division of $m$ by $n$.

(ii) The order of the cyclic subgroup $< g >$ generated by $g$ is $n$.

(iii) For $m \in \mathbb{Z}$, $g^m = e$ if and only if $m$ is a multiple of $n$.

(iv) For $i, j \in \mathbb{Z}$, $g^i = g^j$ if and only if $i - j$ is a multiple of $n$. ●

PROOF

(i) Divide $m$ by $n$ and let $r$ be the remainder. Thus, $0 \leqslant r < n$ and $m = nq + r$ for some integer $q$. (For example, if $n = 5$ and $m = 77$ then $r = 2$ and $q = 15$ and if $n = 5$ and $m = -67$ then $r = 3$ and $q = -14$.) Then

$$g^m = g^{nq+r} = g^{nq} g^r = (g^n)^q g^r = e^q g^r = g^r.$$

(ii) By (i), every power $g^m$ of $g$ is equal to one of the $n$ elements $e, g, \ldots, g^{n-1}$. No two of these $n$ elements are equal, for if $0 \leqslant j < i \leqslant n - 1$ are such that $g^i = g^j$ then $0 < i - j < n$ and $g^{i-j} = e$, contradicting the definition of the order of $g$. Thus, $< g >= \{e, g, \ldots, g^{n-1}\}$ has order $n$.

(iii) If $m$ is a multiple of $n$ then, in (i), $r = 0$ and $g^m = g^0 = e$. If $m$ is not a multiple of $n$ then, by (i), $g^m = g^r$ where $0 < r < n$ and so, by minimality of $n$, $g^m \neq e$.

(iv) $g^i = g^j$ precisely when $g^{i-j} = e$ so this follows from (iii). ●

Combining Theorems 2 and 3(ii), we see that the order of a cyclic subgroup $< g >$ is always equal to the order of $g$. We now use Theorem 3(ii) to get a criterion for a finite group to be cyclic.

## • Theorem 4

Let $G$ be a finite group of order $n$. Then $G$ is cyclic if and only if $G$ has an element of order $n$. ●

PROOF

If $G =< g >$ is cyclic then, by Theorem 3(ii), the order of $g$ must be $n$. Conversely, if $G$ has an element $g$ of order $n$ then, again by Theorem 3(ii), $< g >$ has order $n$ and so must be all of $G$. ●

For example, the group of symmetries of a non-square rectangle has no element of order 4 and so is not cyclic.

**EXERCISE 4**

Which elements of $(\mathbb{R}\backslash\{0\}, \times)$ have finite order?

**EXERCISE 5**

How many cyclic subgroups of order 2 has the dihedral group $D_n$ when $n$ is odd and when $n$ is even?

## 6.4 Orders of Products

The order of a product $ab$ is most easily computed when $ba = ab$. In this case, $(ab)^2 = a^2b^2$ and, in general, $(ab)^n = a^nb^n$. A particular situation where this is useful is in the symmetric group $S_n$. Here, a cycle of length $k$ has order $k$ and every element can be written as a product of disjoint cycles which commute with each other.

**EXERCISE 6**

Find the order of each of the following permutations.

(i) (1 2)(3 4 5);

(ii) (1 2)(3 4);

(iii) (1 2 3 4)(5 6).

Throughout the above exercise, you should have found that the order of the product is the least common multiple of the orders of the factors.

### ● Theorem 5

Let $a, b$ be elements of finite orders $m$ and $n$ respectively in a group $G$. Suppose that $ba = ab$ and that $< a > \cap < b >= \{e\}$ (that is, the only element of $G$ which can be written both as a power of $a$ and as a power of $b$ is $e$). Then the order of $ab$ is the l.c.m. of $m$ and $n$. ●

PROOF

Let $l = mp = nq$ be the l.c.m. of $m$ and $n$. Then

$$(ab)^l = a^lb^l = a^{mp}b^{nq} = (a^m)^p(b^n)^q = e^pe^q = e$$

so the order of $ab$ is at most $l$. Now let $k > 0$ and suppose that $(ab)^k = e$. Then $a^kb^k = e$ so $a^k = b^{-k}$. But then $a^k \in < a > \cap < b >$ so $a^k = e$. Similarly, $b^k = e$. By Theorem 3(iii), $k$ is a common multiple of $m$ and $n$ so $k \geqslant l$. Thus, the order of $ab$ is at least $l$. As we have already seen that it is at most $l$, it must be $l$. ●

### Example 5

Theorem 5 extends to products of more than two commuting factors and, when applied to permutations, tells us that the order of any permutation $f$ is the least common multiple of the lengths of the cycles in the cycle decomposition of $f$. For example, in $S_{10}$, the permutation (1 2 3 4 5 6)(7 8 9 10) has order 12.

## EXERCISE 7

Find the orders of the following permutations in $S_9$:

(i) $\begin{pmatrix} 1 & 2 & 3 & 4 & 5 & 6 & 7 & 8 & 9 \\ 9 & 8 & 7 & 5 & 6 & 4 & 3 & 2 & 1 \end{pmatrix}$;

(ii) $\begin{pmatrix} 1 & 2 & 3 & 4 & 5 & 6 & 7 & 8 & 9 \\ 2 & 3 & 7 & 9 & 5 & 8 & 1 & 6 & 4 \end{pmatrix}$;

(iii) $\begin{pmatrix} 1 & 2 & 3 & 4 & 5 & 6 & 7 & 8 & 9 \\ 2 & 3 & 4 & 5 & 1 & 7 & 8 & 9 & 6 \end{pmatrix}$.

## EXERCISE 8

(This exercise provides examples showing that the conditions $ba = ab$ and $< a > \cap < b > = \{e\}$ in Theorem 5 are necessary.) In each of the following, calculate the order of $ab$ and show that it is not the least common multiple of the orders of the factors $a$ and $b$:

(i) $a = (1\ 2)$, $b = (1\ 2\ 3)$ in the symmetric group $S_3$;

(ii) $a = \text{ref}_0$, $b = \text{rot}_{\frac{2\pi}{7}}$ in the orthogonal group $O_2$;

(iii) $a = (1\ 2\ 3\ 4\ 5\ 6)$, $b = (1\ 4)(2\ 5)(3\ 6)$ in $S_6$.

In (iii), check that $ab = ba$ and that $< a > \cap < b > \neq < e >$.

## 6.5 Orders of Powers

We now consider the order of a power of an element of given order. The next exercise investigates this for order 12 and the theorem which follows it gives the general rule. Before you read the theorem, we suggest that you use your answer to the exercise to predict the general rule.

## EXERCISE 9

If $g \in G$ has order 12, what are the orders of $g^2, g^3, g^5, g^9$?

### Theorem 6

If $g$ has order $n$ then the order of $g^m$ is $\frac{l}{m} = \frac{n}{h}$, where $l$ is the l.c.m. of $m$ and $n$ and $h$ is the h.c.f. of $m$ and $n$. In particular, if $m$ is a factor of $n$ then the order of $g^m$ is $\frac{n}{m}$.

PROOF

The order of $g^m$ is the least positive integer $k$ such that $g^{mk} = e$. By Theorem 3(iii), $g^{mk} = e$ precisely when $mk$ is also a multiple of $n$. Hence, $mk$ is the least common multiple of $m$ and $n$ and so $k = \frac{l}{m}$. The alternative formula $\frac{n}{h}$ is valid because it is always true that $mn = hl$.

For example, if $g$ has order 6 then $g^4$ has order $\frac{12}{4} = 3 = \frac{6}{2}$. In Exercise 9, $g^9$ has order $\frac{36}{9} = 4 = \frac{12}{3}$.

**EXERCISE 10**

(i) Let $G =< g >$ be a cyclic group of order 10. Show that $< g^3 >= G$.

(ii) What is the order of $g^{650}$ when $g$ has order 1000?

## 6.6 Subgroups of Cyclic Groups

We next aim to show that all subgroups of a cyclic group are cyclic. The example and exercise below should help to show some of the ideas which will be used in proving this result.

### Example 6

Let $G =< g >$ be a cyclic group. Suppose that $H$ is a subgroup of $G$ and that $H$ contains $g^4$ but not $g$, $g^2$ or $g^3$. Then $g^{4q} \in H$ for all integers $q$. Can $H$ contain any other powers $g^m$? For example, could $g^{15} \in H$? Now $g^{15} = g^{4\times3+3} = (g^4)^3 g^3$ and so $g^3 = (g^4)^{-3} g^{15}$. Thus, if $g^{15} \in H$ then $g^3 \in H$. As $g^3 \notin H$, $g^{15} \notin H$. Can $g^{-2} \in H$? If so, then $g^2 = (g^{-2})^{-1} \in H$. As $g^2 \notin H$, $g^{-2} \notin H$.

**EXERCISE 11**

Let $G =< g >$ be a cyclic group. Let $H$ be a subgroup of $G$.

(i) Show that if $g^2 \in H$ and $g \notin H$ then $g^7 \notin H$.

(ii) Show that if $g^2 \in H$ and $g \notin H$ then $g^{-3} \notin H$.

### Theorem 7

Let $G$ be a cyclic group and let $H$ be a subgroup of $G$. Then $H$ is cyclic.

PROOF

If $H = \{e\}$ then $H =< e >$ is cyclic. We therefore consider the case when $H \neq \{e\}$.

Thus, $H$ contains $g^t$ for some integer $t \neq 0$. Then $H$ also contains $g^{-t} = (g^t)^{-1}$. As one of $t$ and $-t$ must be positive, $H$ contains $g^s$ for some positive integer $s$ and we can choose the least such positive integer $s$. Thus, $s > 0$, $g^s \in H$ and if $0 < r < s$ then $g^r \notin H$. We now show that $H =< g^s >$. Firstly, $g^s \in H$ and so $< g^s >$ is contained in $H$, that is $(g^s)^q \in H$ for all integers $q$. Can $H$ contain any other elements of $G$? Every element of $G$ has the form $g^m$, $m \in \mathbb{Z}$. Suppose that $m$ is not a multiple of $s$. Then there is a non-zero remainder $r$, $0 < r < s$ when $m$ is divided by $s$. Now $m = sq + r$ for some integer $q$. Then $g^m = (g^s)^q g^r$ and so $g^r = (g^s)^{-q} g^m$. If $g^m \in H$ then, as $(g^s)^{-q}$ is also in $H$, we would have $g^r \in H$. But $0 < r < s$, so we know from above that $g^r \notin H$. Therefore $g^m \notin H$. Hence, the only elements of $H$ are those of the form $g^{sq}$. In other words $H =< g^s >$ is cyclic, generated by $g^s$. ●

Let $G =< g >$ be cyclic of order $n$. Then each power $g^m$, $0 \leqslant m < n$, generates a cyclic subgroup which, by Theorem 6, has order $\frac{n}{h}$ where $h$ is the h.c.f. of $m$ and $n$. In particular, if $m$ is a factor of $n$ then $h = m$ and $| < g > | = \frac{n}{m}$. We appear to have $n$ subgroups $< g^m >$, $0 \leqslant m < n$, but they need not be distinct. For example, we have seen in Example 1 that when $g = \mathrm{rot}_{\frac{2\pi}{5}}$ and $n = 5$, the elements $g$, $g^2$, $g^3$ and $g^4$ all generate the rotation group of the regular pentagon.

## ● Theorem 8

Let $G$ be a finite cyclic group of order $n$ generated by $g$ and let $1 \leqslant m < n$. Then $< g^m >=< g^h >$, where $h$ is the highest common factor of $m$ and $n$. Hence, the number of different subgroups of $G$ is the number, $k$, of different positive factors of $n$. ●

PROOF

Let $d = \frac{n}{h}$. Then, by Theorem 6, $< g^m >$ and $< g^h >$ both have order $d$. But $m = hp$ for some $p$ so $g^m = (g^h)^p \in < g^h >$. Hence, $< g^m > \subseteq < g^h >$ and, as they have the same order, they must be equal.

Let $d_1, d_2, \ldots, d_k$ be a list of the positive divisors of $n$. (For example, if $n = 12$ we have 1, 2, 3, 4, 6, 12.) Then $< g^{d_1} >, < g^{d_2} >, \ldots, < g^{d_k} >$ have different orders $\frac{n}{d_1}, \frac{n}{d_2}, \ldots, \frac{n}{d_k}$ respectively and are $k$ different subgroups of $G$. By Theorem 7, every subgroup $H$ of $G$ has the form $< g^m >$ for some $m$. Then $H =< g^m >=< g^h >$ where $h$ is the h.c.f. of $m$ and $n$. As $h$ is a factor of $n$, $h = d_i$ for some $i$ so $H =< g^{d_i} >$. Hence $< g^{d_1} >, < g^{d_2} >, \ldots, < g^{d_k} >$ are all the subgroups of $G$. ●

## Example 7

Let $G =< g >$ be a cyclic group of order 12. Then the only subgroups of $G$ are $< g >= G$, $< g^2 >$, $< g^3 >$, $< g^4 >$, $< g^6 >$ and $< g^{12} >= \{e\}$ which have orders 12, 6, 4, 3, 2, 1 respectively. Although $< g^8 >$ does not appear explicitly in this list, it is equal to $< g^4 >$. For example, the non-trivial proper subgroups of the rotation group $< \mathrm{rot}_{\frac{\pi}{6}} >$ of the regular dodecagon are the rotation groups of the regular hexagon, the square, the equilateral triangle and the non-square rectangle.

We can also use Theorem 7 to list all the subgroups of an infinite cyclic group such as $(\mathbb{Z}, +)$. These are all cyclic and have the form

$$< m > = < -m > = \{\ldots, -2m, -m, 0, m, 2m, \ldots\} = m\mathbb{Z}.$$

Thus, each subgroup of $(\mathbb{Z}, +)$ consists of the multiples of some integer $m$.

**EXERCISE 12**

How many different subgroups are there in a cyclic group of order 24?

## 6.7 Direct Products of Cyclic Groups

Let $m$ and $n$ be positive integers. Let $G = < a >$ be a cyclic group of order $m$ and $H = < b >$ be a cyclic group of order $n$. Let $m = 2$ and $n = 3$. You can check that $(a, b)$ has order 6 and hence that $G \times H$ is cyclic. This raises the question of whether the direct product of two cyclic groups is always cyclic. The answer is no since, if $m = 2$ and $n = 2$, then none of the four elements of $G \times H$ has order 4 and so $G \times H$ is not cyclic.

● *Theorem 9*

Let $m$ and $n$ be positive integers. Let $G = < a >$ be a cyclic group of order $m$ and $H = < b >$ be a cyclic group of order $n$. Then $G \times H$ is cyclic if and only if $m$ and $n$ are coprime. ●

PROOF

We need to prove two things. Firstly, suppose that $m$ and $n$ are coprime (that is, the h.c.f. of $m$ and $n$ is 1 or, equivalently, their l.c.m. is $mn$). By Theorem 5 applied to $(a, e)$ and $(e, b)$, the order of $(a, b) = (a, e)(e, b)$ is the l.c.m. of $m$ and $n$, which is $mn$. Hence $< (a, b) >$ has order $mn$, the same as that of $G \times H$. By Theorem 4, $G \times H = < (a, b) >$ is cyclic.

Now suppose that $m$ and $n$ are not coprime. The least common multiple $l = ms = nt$ of $m$ and $n$ is then less than $mn$. (For example, if $m = 6$ and $n = 9$ then $l = 18 < 54$.) We have to show that $G \times H$ is not cyclic and to do this we show that no element has order $mn$. Every element of $G \times H$ has the form $(a^x, b^y)$. Now

$$(a^x, b^y)^l = (a^{xl}, b^{yl}) = (a^{mxs}, b^{nyt}) = ((a^m)^{xs}, (b^n)^{yt}) = (e_G, e_H).$$

Thus, the order of $(a^x, b^y)$ is at most $l$ and so is less than $mn$. Therefore, no element of $G \times H$ can have order $mn$ and, by Theorem 4, $G \times H$ is not cyclic. ●

## 6.8 Further Exercises on Chapter 6

**EXERCISE 13**

Let $g$ be an element of a group. Show that $< g > = < g^{-1} >$.

**EXERCISE 14**

Show that the special orthogonal group $SO_2$ has exactly one element of order 2. Is the number of elements of the orthogonal group $O_2$ of order 2 finite or infinite?

## EXERCISE 15

Suppose that the invertible matrix $A$ has finite order in the general linear group $GL_2(\mathbb{R})$. Show that $\det A = \pm 1$.

## EXERCISE 16

What is the least value of $n$ for which the symmetric group $S_n$ has an element of order 12?

## EXERCISE 17

How many elements of order 5 are there in the orthogonal group $O_2$? Investigate which rotations in $O_2$ have finite order.

## EXERCISE 18

How many different subgroups are there in a cyclic group of order 36?

## EXERCISE 19

Let $G =< g >$ be a cyclic group of order 18. How many elements of $G$ have order 18? Investigate the corresponding problem for cyclic groups of different orders.

## EXERCISE 20

Let $p$ be a prime number. Show that a cyclic group of order $p$ has only two subgroups, namely the trivial subgroup and the whole group.

## EXERCISE 21

Let $n$ be a positive integer and consider a pack of $2n$ cards (the most familiar example is when $n = 26$). Number the cards from 1 to $2n$. The **interlacing shuffle** is performed on the pack as follows. Divide the pack into two packs, 1 to $n$ and $n + 1$ to $2n$. Interlace these to form a single pack with card 1 on top followed by card $n + 1$ (the top card from the second pack), card 2, card $n + 2$ and so on. The corresponding permutation when $n = 4$ is

$$\begin{pmatrix} 1 & 2 & 3 & 4 & 5 & 6 & 7 & 8 \\ 1 & 3 & 5 & 7 & 2 & 4 & 6 & 8 \end{pmatrix}$$

and in general it is

$$\begin{pmatrix} 1 & 2 & \dots & n & n+1 & n+2 & \dots & 2n \\ 1 & 3 & \dots & 2n-1 & 2 & 4 & \dots & 2n \end{pmatrix}.$$

Denote this permutation by $p_n$.

(i) Find the orders of $p_4$ and $p_{26}$. (The latter is surprisingly small; this has obvious application for any cardplayer who can perfect the interlacing shuffle.)

(ii) Investigate the order of $p_n$ for other values of $n$.

(iii) A variation on the interlacing shuffle is to interlace with each card from the second pack ($n + 1$ to $2n$) above the corresponding one from the first pack. The corresponding permutation is then

$$q_n = \begin{pmatrix} 1 & 2 & \dots & n & n+1 & n+2 & \dots & 2n \\ 2 & 4 & \dots & 2n & 1 & 3 & \dots & 2n-1 \end{pmatrix}.$$

Investigate the order of $q_n$ for different $n$, in particular $n = 26$.

# 7 • Group Actions

Each element $g$ of the dihedral group $D_4$ permutes the vertices of the square. Given a vertex $v$ of the square there is a vertex, which we might denote by $g * v$, to which $g$ sends $v$. This is an instance of a group acting on a set, in this case the set of vertices of the square. In this chapter, we formalize the notion of a group acting on a set and discuss some particular examples. When we look at a group acting on a set, we hope to gain insight into the symmetry of the set and, at the same time, to obtain a better feel for the group. This is most obvious when the set is a geometrical figure such as the square or circle and the group is its group of symmetries. Another example is where $X = \{1, 2, \ldots, n\}$ and $G$ is the symmetric group $S_n$ of all permutations of $X$. The group $S_4$ also acts on the set of all polynomials in four variables $x_1, x_2, x_3, x_4$ with integer coefficients. This illustrates the fact that a given group may act on different sets. As we shall see later, group actions have an important role in the general theory, not just in the study of particular examples.

## 7.1 Groups Acting on Sets

The basic idea of a group acting on a set $X$ is that each element of $G$ must send each element of $X$ to another element of $X$ in a way that is compatible with the binary operation in $G$. This is expressed formally in two axioms.

We say that a group $G$ **acts** on a non-empty set $X$, or that there is an **action** $*$ of $G$ on $X$ if, for each $g \in G$ and each $x \in X$, there is an element $g * x \in X$ and the following axioms hold:

**GA1:** $e * x = x$ for all $x \in X$;

**GA2:** $g * (h * x) = (gh) * x$ for all $g \in G, h \in G$ and all $x \in X$.

The first axiom says that the effect of the identity element of $G$ is the identity permutation on $X$ and the second says that the combined effect of applying $h$ then $g$ is the same as that of applying $gh$.

### Actions of groups of functions

Several of our basic examples of groups are groups of functions with an identity function id as the neutral element. If $G$ is a group of functions and $X$ is a non-empty set with $f(x) \in X$ for all $x \in X$ and all $f \in G$ then $G$ acts on $X$ by the rule $f * x = f(x)$.

**GA1**: $e_G * x = \mathrm{id} * x = \mathrm{id}(x) = x$ for all $x \in X$.

**GA2**: $g * (h * x) = g * h(x) = g(h(x)) = (gh)(x) = (gh) * x$ for all $g, h \in G$ and all $x \in X$.

Thus, both axioms hold and we do have a group action.

In particular, we have the following examples.

(i) The symmetric group $S_n$ acts on $\{1, 2, \ldots, n\}$ by the rule $f * x = f(x)$. For example $(1\ 4\ 7) * 7 = 1$.

(ii) The orthogonal group $O_2$ and the dihedral groups all act on $\mathbb{R}^2$. For example, $\text{rot}_\pi * (1, 0) = (-1, 0)$.

(iii) Let $X$ be a geometrical figure and let $G$ be its group of symmetries. Then $G$ acts on $X$ with $g * x = g(x)$ for all $g \in G$ and $x \in X$. The rotation group of $X$ also acts on $X$. For example, the orthogonal group $O_2$ and the special orthogonal group $SO_2$ act on the unit circle.

(iv) Let $V = \{1, 2, 3, 4\}$ be the set of vertices of the square, numbered as in Fig 7.1. For all $v \in V$ and $g \in D_4$, $g * v = g(v) \in V$. It follows that $D_4$ acts on $V$ as well as on the square. For example, $r_1 * 1 = 2$.

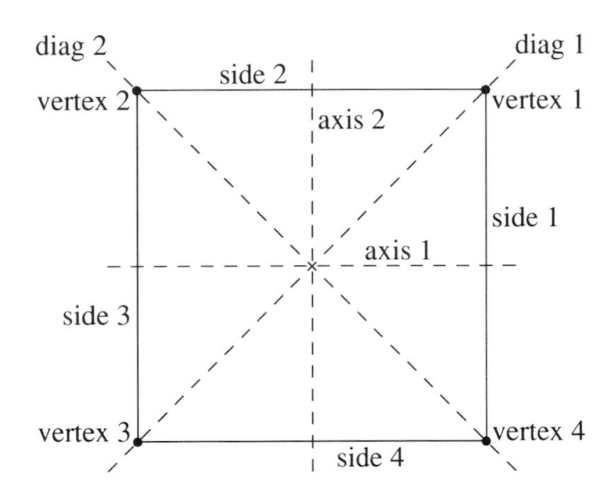


$G = D_4$ acts on

- vertices
  $s_1 : 1 \mapsto 4,\ 2 \mapsto 3,\ 3 \mapsto 2,\ 4 \mapsto 1$
- sides
  $s_1 : 1 \mapsto 1,\ 2 \mapsto 4,\ 3 \mapsto 3,\ 4 \mapsto 2$
- diagonals
  $s_1 : 1 \mapsto 2,\ 2 \mapsto 1$
- axes
  $s_1 : 1 \mapsto 1,\ 2 \mapsto 2$
  $s_1$ =reflection in axis 1

**Fig 7.1** Actions of $D_4$ .

(v) Number the diagonals of the square and consider the two diagonals. Then $D_4$ acts on these by the rule $g * d = g(d)$, where $g(d)$ is the image of $d$ under $g$. For example, $r_1 * 1 = 2$ and $r_2 * 1 = 1$. The action axioms GA1 and GA2 hold because $\text{id}(d) = d$ and $(fg)(d) = f(g(d))$.
Similarly, $D_4$ acts on the four sides of the square and on the two axes.

(vi) The group $D_4$ also acts on the set $N = \{1, 2, 3, 4, 5, 6, 7, 8, 9\}$ of nine small squares. Here, $g * n = g(n)$ is the image of square $n$ under $g$. For example, $s_1 * 2 = 8$.

| 1 | 2 | 3 |
|---|---|---|
| 4 | 5 | 6 |
| 7 | 8 | 9 |

(vii) The orthogonal group $O_2$ acts on the set of all straight lines through the origin, see Fig 1.12.

## Actions of matrix groups

Let $G = GL_2(\mathbb{R})$ or any of its subgroups and let $X = \mathbb{R}^2$. There is an action of $G$ on $X$ given by $A * x = Ax$ for all $A \in G$ and all $x \in X$.
For example, if

$$A = \begin{pmatrix} 1 & 2 \\ 3 & 4 \end{pmatrix} \text{ and } x = \begin{pmatrix} 1 \\ 2 \end{pmatrix} \text{ then } A * x = Ax = \begin{pmatrix} 5 \\ 11 \end{pmatrix}.$$

We check the two group action axioms, remembering that $e_G$ is the identity matrix $I_2$:

**GA1**: $I_2 * x = I_2 x = x$;
**GA2**: by the associativity of matrix multiplication,

$$A * (B * x) = A(Bx) = (AB)x = (AB) * x \text{ for all } A, B \in G \text{ and } x \in X.$$

Thus, $G$ acts on $\mathbb{R}^2$ by matrix multiplication.

## Actions in additive notation

In the case where the group is in additive notation the group action axioms become

**GA1:** $0 * x = x$ for every $x \in X$;
**GA2:** $g * (h * x) = (g + h) * x$ for every $g \in G, h \in G$ and every $x \in X$.

For example, let $G = (\mathbb{R}, +)$ and $X = \mathbb{R}^2$. For $a \in \mathbb{R}$ and $(x, y) \in \mathbb{R}^2$, let

$$a * (x, y) = (x + ay, y).$$

Then $0 * (x, y) = (x, y)$ for all $(x, y) \in \mathbb{R}^2$ so GA1 holds. For GA2,

$$\begin{aligned} a * (b * (x, y)) &= a * (x + by, y) \\ &= (x + by + ay, y) \\ &= (x + (b + a)y, y) \\ &= (b + a) * (x, y) \\ &= (a + b) * (x, y) \text{ for all } a, b \in \mathbb{R} \text{ and for all } (x, y) \in \mathbb{R}^2. \end{aligned}$$

This shows that $*$ is an action of $G$ on $X$. Each element of $\mathbb{R}$ acts by a *shear* parallel to the $x$-axis. This should remind you of the linear transformations which we called shears in Chapter 3. Recall that, for $a \in \mathbb{R}$, the linear transformation determined by the matrix $\begin{pmatrix} 1 & a \\ 0 & 1 \end{pmatrix}$ is a shear. The set of all such matrices is a subgroup $H$ of $GL_2(\mathbb{R})$. In a sense to be made formal in Chapter 9, this group is the same as $(\mathbb{R}, +)$ and the above action of $(\mathbb{R}, +)$ on $\mathbb{R}^2$ is then the same as the one of $H$ by matrix multiplication.

## EXERCISE 1

In each of the following, write down $g * x$ for the given $g$, $x$ and $*$:

(i) $g = (4\ 5\ 7)$, $x = 7$, $*$ is the action of $S_7$ on $\{1, 2, 3, 4, 5, 6, 7\}$;

(ii) $g = \text{rot}_{\frac{\pi}{4}}$, $x = (1, 1)$, $*$ is the action of $O_2$ on $\mathbb{R}^2$;

(iii) $g = s_4 \in D_4$, $x = 4$, $*$ is the action of $D_4$ on the nine small squares;

(iv) $g = \begin{pmatrix} 2 & 7 \\ 1 & 4 \end{pmatrix}$, $x = \begin{pmatrix} 2 \\ -3 \end{pmatrix}$, $*$ is the action of $SL_2(\mathbb{R})$ on $\mathbb{R}^2$ by matrix multiplication.

## 7.2 Orbits

Let $G$ be a group acting on a set $X$. Given $x \in X$, we generate a subset of $X$, called the orbit of $x$, by taking all the elements of $X$ which can be reached from $x$ by applying some element of $G$. This idea was introduced in Chapter 1 when we found the orbit of a small square under the action of $D_4$. Formally, we define the **orbit** of $x$ as

$$\text{orb}(x) = \{y \in X : y = g * x \text{ for some } g \in G\}.$$

### Examples of orbits

We saw several examples of orbits in Chapter 1. The first two examples below are reminders of some of these.

(i) In the action of $D_4$ on the nine small squares, the orbit of 1 is $\{1, 7, 9, 3\}$, the orbit of 2 is $\{2, 4, 8, 6\}$ and the orbit of 5 is just $\{5\}$.

(ii) The orbit of the point $(1, 0)$ in $\mathbb{R}^2$ under the action of the orthogonal group $O_2$ (that is the group of symmetries of the circle) is the unit circle.

(iii) Let $G$ be the group of symmetries of a non-square rectangle. Number the four sides of the rectangle as shown in Fig 7.2. Then $G$ acts on $\{1, 2, 3, 4\}$, for example

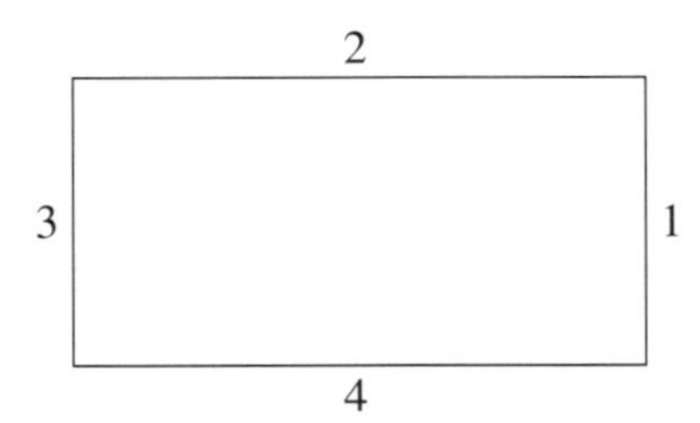


**Fig 7.2** Numbered sides of rectangle.

$r * 1 = 3$ where $r$ is rotation through $\pi$. The orbits are $\{1, 3\}$ and $\{2, 4\}$, that is, the two pairs of opposite sides.

(iv) If we replace the rectangle by a square and take the action of $D_4$ then the four sides are all in the same orbit because, given any two sides of the square there is a symmetry sending the first to the second. In a sense, the number of orbits for a group action on a set measures the symmetry in the set. The fewer the orbits the greater the symmetry.

(v) In the action of $(\mathbb{R}, +)$ on $\mathbb{R}^2$ with the rule $a * (x, y) = (x + ay, y)$, the orbit of the point $(b, c)$ is the set of points $\{(b + ac, c) : a \in \mathbb{R}\}$. If $c \neq 0$ this is the line $y = c$. If $c = 0$ then the orbit is the single point $(b, 0)$. Thus, as shown in Fig 7.3, the orbits of this action consist of lines parallel to the $x$-axis, excluding the $x$-axis itself, together with singleton points on the $x$-axis.

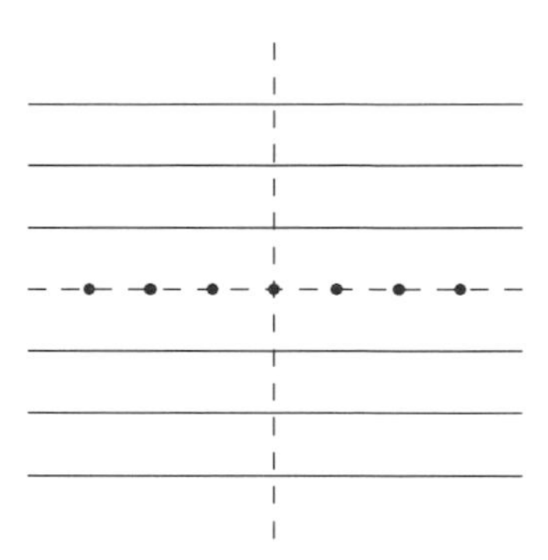

**Fig 7.3** Orbits of the shear action.

(vi) The symmetric group $S_6$ acts on the set $X = \{1, 2, 3, 4, 5, 6\}$ by $g * x = g(x)$. The orbit of each element $i$ of $X$ is the whole set $X$ because we can get any element $j$ of $X$ from $i$ by applying $(i\ j)$ if $j \neq i$ or id if $i = j$.
The cyclic subgroup of $S_6$ generated by $f = (1\ 2\ 3)(4\ 6)$ acts on $X$ by the same rule. For example, $f * 2 = 3$, $f * 4 = 6$, $f * 5 = 5$ and $f^2 * 2 = 1$. The only elements of $X$ that can be obtained from 1 using powers of $f$ are 1, 2 and 3. So these form the orbit of 1. There are three orbits, $\{1, 2, 3\}$, $\{4, 6\}$ and $\{5\}$, one for each of the cycles in the cycle decomposition of $f$.

## EXERCISE 2

Divide the square into 16 numbered squares. How many different orbits are there for the action of $D_4$ on $\{1, 2, \ldots, 16\}$?

## EXERCISE 3

The cyclic subgroup of $S_7$ generated by $(1\ 2)(3\ 4\ 5\ 7)$ acts on the set $\{1, 2, 3, 4, 5, 6, 7\}$ by the rule $g * x = g(x)$. What are the orbits?

## 7.3 Stabilizers

Let $G$ be a group acting on a non-empty set $X$ and let $x \in X$. The stabilizer of $x$ consists of those elements of $G$ which leave $x$ unchanged. This idea was introduced informally in Chapters 1 and 2 where several examples were given, for instance, on pages 5 and 13. Formally, the **stabilizer** of $x$ is the set

$$\text{stab}(x) = \{g \in G : g * x = x\}.$$

● *Theorem 1*

Let $G$ be a group acting on a non-empty set $X$ and let $x \in X$. Then stab($x$) is a subgroup of $G$. ●

PROOF

We use the subgroup criterion.

**SG1:** Since $e * x = x$ by GA1, we have that $e \in \text{stab}(x)$ and so stab($x$) is non-empty.

**SG2:** Let $g, h \in \text{stab}(x)$. Then $g * x = x$ and $h * x = x$. Now $(gh) * x = g * (h * x) = g * x = x$ using GA2. Thus, $gh \in \text{stab}(x)$ as required.

**SG3:** Let $g \in \text{stab}(x)$. Then $x = g * x$ and so $g^{-1} * x = g^{-1} * (g * x) = (g^{-1}g) * x = e * x = x$. Hence, $g^{-1} \in \text{stab}(x)$ as required.

By the subgroup criterion, stab($x$) is a subgroup of $G$. ●

## Examples of stabilizers

(i) When the dihedral group $D_4$ acts on the nine squares, $\text{stab}(2) = \{e, s_3\} = \{\text{rot}_0, \text{ref}_\pi\}$. This is also the stabilizer of the point (0, 1) when the orthogonal group $O_2$ acts on the unit circle.

(ii) The symmetric group $S_4$ acts on $\{1, 2, 3, 4\}$ by $g * x = g(x)$. The stabilizer of 1 is the set of permutations which leave 1 unchanged that is

$$\text{stab}(1) = \{\text{id}, (2\ 3), (3\ 4), (2\ 4), (2\ 3\ 4), (2\ 4\ 3)\}.$$

(iii) The group $O_2$ acts on the set of all figures $A$ in the Euclidean plane by the rule $f * A = f(A)$. Thus, $f * A = A$ means $f(A) = A$ and so the stabilizer of each figure is its group of symmetries.

**EXERCISE 4**

The dihedral group $D_3$ acts on the four numbered equilateral triangles shown.

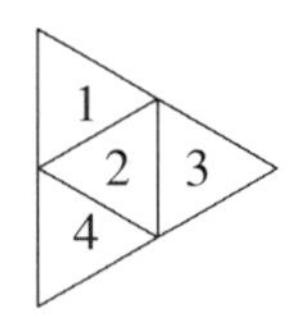


For each of the four triangles, find its stabilizer under the action of $D_3$.

## 7.4 Permutations Arising from Group Actions

When a group $G$ acts on a set $X$, to each element $g$ of $G$ there is associated a function $f_g : X \to X$ given by the rule $f_g(x) = g * x$ for all $x \in X$. Axiom GA1 says that $f_e = \text{id}_X$ and GA2 says that $f_g f_h = f_{gh}$ for all $g, h \in G$. In particular,

$f_g f_{g^{-1}} = f_e = \mathrm{id}_X$ and $f_{g^{-1}} f_g = f_e = \mathrm{id}_X$. Hence, $f_g$ has inverse $f_{g^{-1}}$ and so is a bijective function from $X$ to $X$; that is, a permutation of $X$.

In the action of $D_4$ on the vertices of the square, the eight permutations $f_g$ of $V$ corresponding to the eight elements $g$ of $D_4$ are listed on page 5, for example,

$$f_{r_1} = \begin{pmatrix} 1 & 2 & 3 & 4 \\ 2 & 3 & 4 & 1 \end{pmatrix}.$$

For $A \in GL_2(\mathbb{R})$ the permutation $f_A : \mathbb{R}^2 \to \mathbb{R}^2$ arising from the action of $GL_2(\mathbb{R})$ on $\mathbb{R}^2$ is given by $f_A(x) = A * x = Ax$. This agrees with the use of the notation $f_A$ introduced in Chapter 3 for the linear transformation $f_A : \mathbb{R}^2 \to \mathbb{R}^2$. This was defined by the same rule, $f_A(x) = Ax$ for all $x \in \mathbb{R}^2$.

**EXERCISE 5**

The dihedral group $D_3$ acts on the four numbered equilateral triangles shown in Exercise 4. For each of the six elements $g$ of $D_3$, write down the corresponding permutation $f_g$ in $S_4$.

## 7.5 The Alternating Group

This section introduces an important subgroup of the symmetric group $S_n$. This subgroup, which will be called the alternating group, arises as the stabilizer of a polynomial in $n$ variables under an action of $S_n$. In the case $n = 4$, this is the action which we investigated in Section 2.1.

The symmetric group $S_n$ acts on the set of polynomials in $n$ variables $x_1, x_2, \ldots, x_n$ with integer coefficients by

$$f * p(x_1, x_2, \ldots, x_n) = p(x_{f(1)}, x_{f(2)}, \ldots, x_{f(n)})$$

where $p = p(x_1, x_2, \ldots, x_n)$ is a polynomial in $x_1, x_2, \ldots, x_n$ and $f \in S_n$. For example, if $n = 4$, $f = (1\ 2\ 4)$ and $p = x_1^2 x_2 - x_3 x_4$ then $f * p = x_2^2 x_4 - x_3 x_1$.

We now check that this is a group action.

$$\begin{aligned} \textbf{GA1:} \quad \mathrm{id} * p(x_1, x_2, \ldots, x_n) &= p(x_1, x_2, \ldots, x_n). \\ \textbf{GA2:} \quad g * (f * p(x_1, x_2, \ldots, x_n)) &= g * p(x_{f(1)}, x_{f(2)}, \ldots, x_{f(n)}) \\ &= p(x_{g(f(1))}, x_{g(f(2))}, \ldots, x_{g(f(n))}) \\ &= gf * p(x_1, x_2, \ldots, x_n). \end{aligned}$$

We should point out that the action respects the addition and multiplication of the polynomials. That is, for all polynomials $p$ and $q$ in $x_1, x_2, \ldots, x_n$,

$$f * (p + q) = f * p + f * q \text{ and } f * (pq) = (f * p)(f * q).$$

For example,

$$(1\ 2\ 3) * (x_1 - x_2)(x_2 - x_3) = (x_2 - x_3)(x_3 - x_1).$$

Consider now the polynomial

$$a = (x_1 - x_2)(x_1 - x_3)(x_1 - x_4)(x_2 - x_3)(x_2 - x_4)(x_3 - x_4).$$

When a permutation is applied to $a$, each of the six factors of $a$ is sent to plus or minus one of the six factors. For example $(2\ 4)$ sends the first factor to the third factor and the fourth factor to minus the sixth factor. Hence, $a$ will be sent to $\pm a$, depending on the number of minuses. Thus, $a$ has just two elements in its orbit, $a$ and $-a$. This is a special polynomial called the alternating polynomial. In general, for $n \geqslant 2$, the **alternating polynomial** in $n$ variables is defined as

$$a_n = a_n(x_1, x_2, \ldots, x_n) = \prod_{i<j}(x_i - x_j).$$

That is, $a_n$ is the product of all the polynomials of the form $x_i - x_j$ with $i < j$. For example, if $n = 4$ then $a_n$ is the above polynomial $a$.

As different permutations are applied to $a_n$ only two polynomials, $a_n$ and $-a_n$ are obtained. This divides $S_n$ into two subsets: those that send $a_n$ to $a_n$, in other words the stabilizer of $a_n$, and those that send $a_n$ to $-a_n$. We will use this to answer the problem about products of even and odd numbers of transpositions posed in our discussion of the 15-puzzle in Section 2.7.

The stabilizer of $a_n$ is called the **alternating group** $A_n$. To decide whether a given permutation $f$ is in $A_n$, we use adjacent transpositions $(i\ i+1)$. If we act on $a_n$ by $(i\ i+1)$ we obtain $-a_n$. This is most easily seen if we write

$$\begin{array}{rcl} a_n = (x_1 - x_2)(x_1 - x_3)(x_1 - x_4) & \ldots & (x_1 - x_n) \\ (x_2 - x_3)(x_2 - x_4) & \ldots & (x_2 - x_n) \\ & \ddots & \vdots \\ & & (x_{n-1} - x_n). \end{array}$$

Applying $t = (i\ i+1)$ to $a_n$ only affects the $i$th and $(i+1)$th rows in this display of $a_n$. In any of the other rows it simply swops two factors. Rows $i$ and $i+1$ change from

$$\begin{array}{rcl} (x_i - x_{i+1})(x_i - x_{i+2})(x_i - x_{i+3}) & \ldots & (x_i - x_n) \\ (x_{i+1} - x_{i+2})(x_{i+1} - x_{i+3}) & \ldots & (x_{i+1} - x_n) \end{array}$$

into

$$\begin{array}{rcl} (x_{i+1} - x_i)(x_{i+1} - x_{i+2})(x_{i+1} - x_{i+3}) & \ldots & (x_{i+1} - x_n) \\ (x_i - x_{i+2})(x_i - x_{i+3}) & \ldots & (x_i - x_n). \end{array}$$

The first term in row $i$ has been multiplied by $-1$ and the other terms have been switched between the two rows without change of sign. Thus, $t * a_n = -a_n$ and so $t \notin A_n$ as it does not stabilize $a_n$. Moreover, $t * (-a_n) = -(-a_n) = a_n$ so $t$ interchanges $a_n$ and $-a_n$.

Now consider a product $f$ of two adjacent transpositions $t_1t_2$. Then

$$f * a_n = (t_1t_2) * a_n = t_1 * (t_2 * a_n) = t_1 * (-a_n) = a_n$$

so $f \in A_n$ as it does stabilize $a_n$.



If $f$ is a product of three adjacent transpositions then $f * a_n = -a_n$ and so on. In general, if $f$ is a product of an odd number of adjacent transpositions then $f * a_n = -a_n$ and if $f$ is a product of an even number of adjacent transpositions then $f * a_n = a_n$. Next, consider an arbitrary transposition $t = (i\ j)$. As $(i\ j) = (j\ i)$, we may assume $i < j$. In Chapter 2, a single transposition was written as a product of adjacent transpositions. The formula is

$$(i\ j) = (i\ i+1)(i+1\ i+2)\dots(j-2\ j-1)(j-1\ j)(j-2\ j-1)$$
$$\dots(i+1\ i+2)(i\ i+1).$$

There is an odd number $2(j-i)-1$, of adjacent transpositions on the right-hand side so $t$ is a product of an odd number of adjacent transpositions. Hence, $t * a_n = -a_n$ for any transposition $t$. Repeating the above argument with arbitrary transpositions rather than adjacent ones shows that

- if $f$ is a product of an odd number of transpositions then $f * a_n = -a_n$,
- if $f$ is a product of an even number of transpositions then $f * a_n = a_n$.

We can draw the following conclusions.

- $f * a_n$ must be $a_n$ or $-a_n$ so the orbit of $a_n$ is $\{a_n, -a_n\}$.
- A permutation $f$ cannot be both a product of an even number of transpositions and a product of an odd number of transpositions.
- In particular, a single transposition cannot be written as a product of an even number of transpositions. Hence the 15-puzzle as posed in Section 2.7 is impossible.

A permutation $f$ in $S_n$, $n \geqslant 2$, is said to be **even** (respectively **odd**) if it can be written as a product of an even (respectively odd) number of transpositions.

- Every permutation in $S_n$, $n \geqslant 2$, is even or odd but cannot be both.
- $\mathrm{id} = (1\ 2)(1\ 2)$ is even and transpositions are odd.
- The alternating group $A_n$, that is, the stabilizer of $a_n$, consists of the even permutations. The odd permutations send $a_n$ to $-a_n$.
- The formula

  $$(a_1\ a_2\ a_3\ \dots\ a_k) = (a_1\ a_2)(a_2\ a_3)\dots(a_{k-1}\ a_k)$$

  in Chapter 2 for a cycle of length $k$ as a product of $k-1$ transpositions shows that, somewhat perversely, cycles of even length are odd and cycles of odd length are even.
- If $f$ is a product of $k$ transpositions and $g$ is a product of $l$ transpositions then $fg$ is a product of $k+l$ transpositions. Hence, the rules for combining even and odd permutations by composition are the same as for combining even and odd integers by *addition*. They are summarized in Table 7.1 which looks like the Cayley table of a group with two elements, odd and even. Indeed it is. We shall use it later to illustrate the important idea of factor group.

The **sign** of a permutation $f$, written $\mathrm{sgn}\ f$, is defined to be $+1$ if $f$ is even and $-1$ if $f$ is odd.

- The rules in Table 7.1 can be summarized in the single equation

  $$\mathrm{sgn}\,(fg) = \mathrm{sgn}\ f\ \mathrm{sgn}\ g.$$

| | even | odd |
|---|---|---|
| even | even | odd |
| odd | odd | even |

**Table 7.1** Rules for parity.

In the example we used in Section 2.5 to illustrate the cycle decomposition algorithm

$$f = (1\ 8\ 3)(2\ 6)(4\ 9\ 12\ 11)(5\ 10) = (1\ 8)(8\ 3)(2\ 6)(4\ 9)(9\ 12)(12\ 11)(5\ 10)$$

so $f$ is odd, $\operatorname{sgn} f = -1$ and $f \notin A_{12}$.

Let $f \in S_n$ have cycle decomposition $f = f_1 f_2 \ldots f_s$ and, for $1 \leqslant i \leqslant s$, let $k_i$ be the length of $f_i$. For example, with $f$ as above, $s = 4$, $k_1 = 3$, $k_2 = 2$, $k_3 = 4$ and $k_4 = 2$. We know that $\operatorname{sgn} f_i = (-1)^{k_i - 1}$ for each $i$. By repeated use of the formula $\operatorname{sgn}(fg) = \operatorname{sgn} f \operatorname{sgn} g$, we deduce that

$$\operatorname{sgn} f = (-1)^{(k_1-1)+(k_2-1)+\ldots+(k_s-1)}.$$

This is sometimes taken as the definition of $\operatorname{sgn} f$. In the example, $\operatorname{sgn} f = (-1)^{2+1+3+1} = -1$.

In $S_3$, the three transpositions $(1\ 2)$, $(1\ 3)$ and $(2\ 3)$ are odd and the other three elements id, $(1\ 2\ 3)$ and $(1\ 3\ 2)$ are even. Thus, the alternating group $A_3$ has order 3. This is a special case of the next theorem.

## Theorem 2

The alternating group $A_n$ has order $\frac{1}{2}n!$.

PROOF

Let $B_n$ denote the set of all odd permutations in $S_n$. Every permutation in $S_n$ is in $A_n$ or $B_n$ but not in both. So $n! = |S_n| = |A_n| + |B_n|$. Let $k = |A_n|$ and let $m = |B_n| = n! - k$. List the $k$ even permutations as $A_n = \{f_1, f_2, \ldots, f_k\}$ and the $m$ odd permutations as $B_n = \{g_1, g_2, \ldots, g_m\}$. Let $t$ be the transposition $(1\ 2)$. Then $f_1 t, f_2 t, \ldots, f_k t$ are $k$ odd permutations. They are distinct for if $f_i t = f_j t$ then, by cancellation of $t$, $f_i = f_j$. Thus, $m = |B_n| \geqslant k$. Similarly, $g_1 t, g_2 t, \ldots, g_m t$ are $m$ distinct elements of $A_n$ and $k \geqslant m$. Therefore, $k = m$ and so $|A_n| = k = \frac{n!}{2}$.

## EXERCISE 6

For the permutation $f = \begin{pmatrix} 1 & 2 & 3 & 4 & 5 & 6 & 7 & 8 & 9 \\ 5 & 2 & 9 & 1 & 7 & 3 & 4 & 6 & 8 \end{pmatrix}$, determine whether $f$ is odd or even, write down $\operatorname{sgn} f$ and state whether $f \in A_9$.

## EXERCISE 7

Show that $\operatorname{sgn}(f^2) = +1$ for all permutations $f$. Deduce that there is no permutation $f \in S_6$ such that $f^2 = (1\ 2\ 3\ 4\ 5\ 6)$.

## 7.6 Further Exercises on Chapter 7

### EXERCISE 8

Show that the group $(\mathbb{R}\backslash\{0\}, \times)$ acts on the set $\mathbb{R}^2$ by the rule $r * (x, y) = (rx, ry)$.

### EXERCISE 9

Show that the group $(\mathbb{Z}, +)$ acts on the set $\mathbb{Q}$ of all rational numbers by the rule $z * q = z + q$. What is the orbit of $2\frac{3}{4}$?

In the next three exercises, the orbits are subsets of $\mathbb{R}^2$. We have seen circles arising as orbits in $\mathbb{R}^2$. In these exercises the orbits include straight lines, ellipses and hyperbolas. If you have access to a graphics calculator with the capability of plotting curves given in parametric form, or to a computer with graphics, we encourage you to use it as indicated in the first of these exercises.

### EXERCISE 10

Fix a non-zero real number $t$ and let $H$ be the subgroup $\left\{\begin{pmatrix} \cos\theta & -t\sin\theta \\ \frac{1}{t}\sin\theta & \cos\theta \end{pmatrix} : \theta \in \mathbb{R}\right\}$ of $GL_2(\mathbb{R})$. Then $H$ acts on $\mathbb{R}^2$ by matrix multiplication. Thus

$$\begin{pmatrix} \cos\theta & -t\sin\theta \\ \frac{1}{t}\sin\theta & \cos\theta \end{pmatrix} * \begin{pmatrix} x \\ y \end{pmatrix} = \begin{pmatrix} x\cos\theta - yt\sin\theta \\ \frac{x}{t}\sin\theta + y\cos\theta \end{pmatrix}.$$

Investigate the orbits of this action.

For example, when $t = 2$, the orbit of $P = \begin{pmatrix} 2 \\ 0 \end{pmatrix}$ consists of all points of the form $\begin{pmatrix} 2\cos\theta \\ \sin\theta \end{pmatrix}$. Plotting such points, we obtain an ellipse.

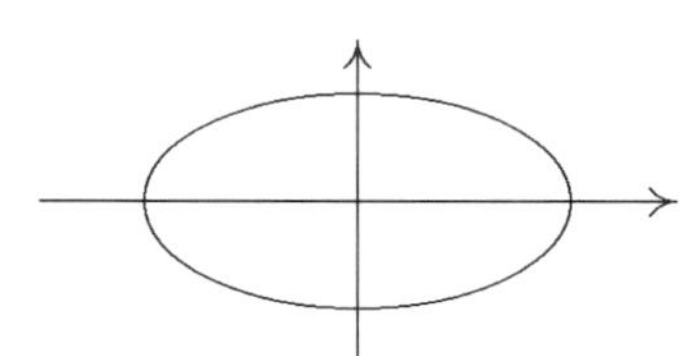

### EXERCISE 11

Let $H$ be the subgroup $\left\{\begin{pmatrix} a & b \\ b & a \end{pmatrix} : a, b \in \mathbb{R} \text{ and } a^2 - b^2 = 1\right\}$ of $GL_2(\mathbb{R})$. Then $H$ acts on $\mathbb{R}^2$ by matrix multiplication. Thus

$$\begin{pmatrix} a & b \\ b & a \end{pmatrix} * \begin{pmatrix} x \\ y \end{pmatrix} = \begin{pmatrix} ax + by \\ bx + ay \end{pmatrix}.$$

Investigate the orbits of this action.

## EXERCISE 12

The group $(\mathbb{R}\backslash\{0\}, \times)$ acts on the set $\mathbb{R}^2$ by $r * (x, y) = (rx, ry)$. Find $\text{orb}(1, 0)$, $\text{orb}(0, 0)$ and $\text{orb}(1, 1)$. Hence describe all the orbits of this action.

## EXERCISE 13

Show that, for the action of the general linear group $GL_2(\mathbb{R})$ on $\mathbb{R}^2$ by matrix multiplication, the orbit of $\begin{pmatrix} 1 \\ 0 \end{pmatrix}$ consists of all points of $\mathbb{R}^2$ except $\begin{pmatrix} 0 \\ 0 \end{pmatrix}$.

## EXERCISE 14

Show that, for the action of the group $GL_2(\mathbb{R})$ of all $2 \times 2$ invertible real matrices on $\mathbb{R}^2$ by matrix multiplication, the stabilizer of the point $P = \begin{pmatrix} 1 \\ 1 \end{pmatrix}$ is the set of all $2 \times 2$ matrices of the form $\begin{pmatrix} a & 1-a \\ c & 1-c \end{pmatrix}$ where $a, c \in \mathbb{R}$ and $a \neq c$. Find the stabilizer of $P$ for the action of the special linear group $SL_2(\mathbb{R})$, that is the group of $2 \times 2$ matrices of determinant 1.

## EXERCISE 15

The cyclic subgroup $G = \langle f \rangle$ of $S_8$ generated by $f = (1\ 2\ 3)(5\ 6\ 7\ 8)$ acts on $\{1, 2, 3, 4, 5, 6, 7, 8\}$ by $g * x = g(x)$. Find the orbits and stabilizers of 1, 4 and 5.

## EXERCISE 16

Consider the action of $D_4$ on the nine small squares. For each of the eight elements $g$ of $D_4$, write down the corresponding permutation $f_g$ in $S_9$.

## EXERCISE 17

In this question you are asked to investigate orbits and stabilizers of the action of $S_4$ on polynomials in $x_1, x_2, x_3, x_4$. The polynomials are chosen to give a variety of stabilizers. You will save time if you observe patterns in the polynomials.

(i) Find $|\text{orb}(p)|$ and $|\text{stab}(p)|$ for each of the following polynomials $p$.

(a) $p = x_1^2 + x_2^2 + x_3^2 + x_4$;

(b) $p = x_1x_2^2 + x_2x_3^2 + x_3x_1^2 + x_4$;

(c) $p = x_1x_2^2x_3^3 + x_2x_3^2x_4^3 + x_3x_4^2x_1^3 + x_4x_1^2x_2^3$;

(d) $p = x_1x_2 + x_2x_3 + x_3x_4 + x_4x_1$.

(ii) Can you suggest a relationship between $|S_4|$, $|\text{orb}(p)|$ and $|\text{stab}(p)|$? Experiment with a few polynomials of your own choosing.

## EXERCISE 18

For $n \geqslant 2$, let

$$f_n = \begin{pmatrix} 1 & 2 & \dots & \dots & n \\ n & n-1 & \dots & \dots & 1 \end{pmatrix} .$$

Is it true that $f_n$ is even whenever $n$ is even and odd whenever $n$ is odd? If not, then write down the rule for the parity (odd or even) of $f_n$ when $n = 2m$ is even and when $n = 2m + 1$ is odd.

## EXERCISE 19

(i) Write each of the following permutations as a product of transpositions of the form $(1\ a), a \neq 1$:

(a) the cycle $(1\ i\ j)$ where $1, i, j$ are distinct;

(b) the permutation $f = (1\ 4)(2\ 3)(5\ 8)(6\ 7)$.

(ii) Use the formula given by (i)(a) to write the permutation $f$ in (i)(b) as a product of cycles of the form given in (i)(a).

(iii) Show that if $n > 2$ then every even permutation can be written as a product of 3-cycles. Are all products of 3-cycles even?

## EXERCISE 20

Write down all the elements of the alternating group $A_4$.

# 8 • Equivalence Relations and Modular Arithmetic

Often in mathematics, there are sets in which some of the elements are in some sense 'similar' or 'related'. In $\mathbb{R}^2$, any two points on the same circle with its centre at the origin are at the same distance from the origin and so might be regarding as being 'similar'. In the group of symmetries of the square, the two reflections in the diagonals are 'similar' to each other but are in some respects different to the other two reflections. The idea of an equivalence relation formalizes this idea of similarity. An equivalence relation on a set 'partitions' the set into non-overlapping subsets, which consist of elements similar to each other. We shall see many equivalence relations arising naturally in the context of group actions. In the final section of the chapter we take the equivalence classes of one particular equivalence relation and define two binary operations, addition and multiplication modulo $n$. This gives rise to further examples of groups. In Chapter 10, we will apply general group theoretical methods to these groups and obtain some theorems of elementary number theory.

## 8.1 Partitions

There are many instances in mathematics where there is a natural way of dividing a set into non-overlapping subsets. This gives rise to the idea of a partition. For example, when $D_4$ acts on the nine numbered squares, the squares divide into three orbits:

| 1 | 2 | 3 |
|---|---|---|
| 4 | 5 | 6 |
| 7 | 8 | 9 |

| | |
|---|---|
| the corners | $\{1, 3, 7, 9\}$ |
| the middles of sides | $\{2, 4, 6, 8\}$ |
| the centre | $\{5\}$. |

We can also divide the group $D_4$ into non-overlapping subsets by considering the orbit of $1, \{1, 3, 7, 9\}$:

| | |
|---|---|
| elements of $D_4$ sending 1 to 1: | $\{e, s_4\}$ |
| elements of $D_4$ sending 1 to 3: | $\{r_3, s_3\}$ |
| elements of $D_4$ sending 1 to 7: | $\{r_1, s_1\}$ |
| elements of $D_4$ sending 1 to 9: | $\{r_2, s_2\}$. |

These are two examples of partitions and we shall see that similar partitions arise from any group action.

A **partition** of a non-empty set $A$ is given by a collection of non-empty subsets $A_i$ of $A$ such that $A$ is the union $(\cup A_i)$ of the subsets $A_i$ (that is, each element of $A$ belongs to at least one $A_i$) and $A_i \cap A_j = \emptyset$ whenever $A_i \neq A_j$ (that is, the subsets $A_i$ do not overlap).

### Examples of partitions

(i) The set of all humans has a natural partition into two subsets, those that are female and those that are male.

(ii) Let $A = \mathbb{R}^2$. Measuring distances from the origin gives a partition of $A$ into subsets, all but one of which is a circle with its centre at the origin. The exception consists of the single point at the origin. Notice that these subsets are again orbits of a group action, namely that of the orthogonal group $O_2$ on $\mathbb{R}^2$.

(iii) Consider the set $\mathbb{Z}$ of all integers. Taking remainders on division by 5 gives a partition of $\mathbb{Z}$ into the following five subsets:

$$\begin{array}{ll} \text{remainder } 0: & \{\ldots, -15, -10, -5, 0, 5, 10, 15, \ldots\} \\ \text{remainder } 1: & \{\ldots, -14, -9, -4, 1, 6, 11, 16, \ldots\} \\ \text{remainder } 2: & \{\ldots, -13, -8, -3, 2, 7, 12, 17, \ldots\} \\ \text{remainder } 3: & \{\ldots, -12, -7, -2, 3, 8, 13, 18, \ldots\} \\ \text{remainder } 4: & \{\ldots, -11, -6, -1, 4, 9, 14, 19, \ldots\}. \end{array}$$

### EXERCISE 1

Use the orbit of square number 2 to obtain a partition of $D_4$ as was done above using the orbit of 1.

### EXERCISE 2

Use division by 4 to obtain a partition of $\mathbb{Z}$ into four subsets.

## 8.2 Relations

We have indicated above that there is a connection between orbits of group actions and partitions. Before explaining this in general we discuss the idea of an equivalence relation which essentially gives an alternative approach to partitions.

To specify a relation we need to give a rule for when two elements of a given set are to be regarded as 'related'. The formal definition, given below, may look a bit forbidding but we shall explain how it can be interpreted in the above way. If $A$ is a non-empty set, then the Cartesian product $A \times A$ is the set of all ordered pairs $(a, b)$ with $a \in A$ and $b \in A$. For example, $\mathbb{R} \times \mathbb{R}$ is the coordinate plane.

A **relation** $R$ on a non-empty set $A$ is a non-empty subset of the Cartesian product $A \times A$.

Rather than specifying a subset $R$ of $A \times A$, it is usual to give a rule for when $(a, b) \in R$. We write $aRb$ rather than $(a, b) \in R$ and read this as '$a$ is related to $b$ (under the relation $R$)'.

### Examples of relations

(i) Let $X$ be the set of all humans. We could define a relation $R$ on $X$ by saying that, for $a, b \in X$, $aRb$ precisely when $a$ and $b$ have a grandparent in common.

(ii) We can specify a relation $R$ on the set $X$ of all species of birds by the rule that $aRb$, if a bird of species $a$ has eaten one of species $b$. Thus, for example, *peregrine* $R$ *pigeon* but, as far as we know, *blue tit* is not related to *ptarmigan* under $R$.

(iii) There is a relation $\geqslant$ ('greater than or equal to') on $\mathbb{R}$ for which it is true that $2 \geqslant 2$ and $2 \geqslant 1$ but false that $\pi \geqslant 4$.

(iv) Let $n \geqslant 2$ be an integer. We define a relation called **equivalence modulo** $n$ on the set $\mathbb{Z}$ of all integers as follows:

for $a, b \in \mathbb{Z}$, $aRb$ if and only if $a - b$ is divisible by $n$.

For this relation, $aRb$ is written $a \equiv b \bmod n$. For example, $17 \equiv 2 \bmod 5$ and $-39 \equiv 3 \bmod 7$.

We shall be concerned with relations which have three particular properties. A relation $R$ on a non-empty set $A$ is said to be

- **reflexive** if $aRa$ for all $a \in A$;
- **symmetric** if, whenever $aRb$, we also have $bRa$;
- **transitive** if, whenever $aRb$ and $bRc$, we also have $aRc$.

A relation $R$ is an **equivalence relation** if it is reflexive, symmetric and transitive.

The 'common grandparent' relation $R$ is certainly reflexive and symmetric but not transitive. We are confident that the 'birds eating birds' relation is not reflexive (no hawfinch has ever eaten another hawfinch) and not symmetric (pigeons do not eat peregrines). We are not sure about transitivity: eagle owls eat goshawks and goshawks eat sparrowhawks but do eagle owls eat sparrowhawks?

The 'greater than or equal to' relation $\geqslant$ is reflexive and transitive but it is not symmetric because $2 \geqslant 1$ whereas $1 \not\geqslant 2$.

Now consider equivalence modulo $n$. The use of the word 'equivalence' in the name of this relation suggests that it is an equivalence relation. We check for reflexivity, symmetry and transitivity in turn.

- Reflexivity: let $a \in \mathbb{Z}$. Then $a - a = 0 = 0 \times n$ so $a \equiv a \bmod n$.
- Symmetry: let $a, b \in \mathbb{Z}$ be such that $a \equiv b \bmod n$, that is $a - b = mn$ for some integer $m$. Then $b - a = -mn = (-m)n$ is divisible by $n$ so $b \equiv a \bmod n$.
- Transitivity: let $a, b, c \in \mathbb{Z}$ be such that $a \equiv b \bmod n$ and $b \equiv c \bmod n$, that is $a - b = mn$ and $b - c = pn$ for some integers $m$ and $p$. Then $a - c = (a - b) + (b - c) = mn + pn = (m + p)n$ so $a \equiv c \bmod n$.

Thus, equivalence modulo $n$ is an equivalence relation.

## EXERCISE 3

We define a relation $R$ on the set $\mathbb{Q}$ of all rational numbers as follows:

for $a, b \in \mathbb{Q}$, $aRb \Leftrightarrow a - b \in \mathbb{Z}$.

For example, $1R1$, $2R5$, $1\frac{1}{5}R3\frac{1}{5}$ but 1 is not related to $\frac{3}{4}$. Show that $R$ is an equivalence relation on $\mathbb{Q}$.

## 8.3 Equivalence Classes

In this section, we describe how an equivalence relation on a set $A$ gives rise to a partition of $A$.

Let $A$ be a non-empty set with an equivalence relation $R$ and let $a \in A$. The **equivalence class** of $a$, written $\overline{a}$, is the set

$$\{b \in A : bRa\}$$

of all elements of $A$ related to $a$.

### Example 1

If $A = \mathbb{Z}$ and $R$ is equivalence modulo 5 then

$$\overline{0} = \{\ldots, -15, -10, -5, 0, 5, 10, 15, \ldots\},$$

$$\overline{1} = \{\ldots, -14, -9, -4, 1, 6, 11, 16, \ldots\},$$

$$\overline{2} = \{\ldots, -13, -8, -3, 2, 7, 12, 17, \ldots\},$$

$$\overline{3} = \{\ldots, -12, -7, -2, 3, 8, 13, 18, \ldots\},$$

$$\overline{4} = \{\ldots, -11, -6, -1, 4, 9, 14, 19, \ldots\}.$$

For other elements of $\mathbb{Z}$ the equivalence classes repeat these, for example, $\overline{5} = \overline{0}$, $\overline{6} = \overline{1}$ and $\overline{-1} = \overline{4}$. These equivalence classes are precisely the sets in the partition of $\mathbb{Z}$ we obtained in Section 8.1 by taking remainders on division by 5.

There are several features of this example which are typical of what happens in general.

- Each element $a$ is in its own equivalence class $\overline{a}$. This is true for an arbitrary equivalence relation $R$ on a non-empty set $A$ because of reflexivity whereby $aRa$ and hence $a \in \overline{a}$ for all $a \in A$. This gives us the first property for a partition, $A$ is the union of the equivalence classes.
- Sometimes $\overline{a} = \overline{b}$ when $a \neq b$. For example, $\overline{6} = \overline{1}$. We shall need to know precisely when this happens. The following simple but important theorem tells us when it does and is illustrated by the fact that, in the quoted example, $6 \equiv 1 \bmod 5$.

### Theorem 1

Let $R$ be an equivalence relation on a non-empty set $A$ and let $a, b \in A$. Then $\overline{a} = \overline{b}$ if and only if $aRb$.

PROOF

For the 'if' part, suppose that $aRb$. We have to show that the sets $\overline{a}$ and $\overline{b}$ are equal. To do this, first take an arbitrary element $c \in \overline{a}$. Thus, $cRa$. We have $cRa$ and $aRb$ so, by transitivity, $cRb$. Therefore, $c \in \overline{b}$ for all $c \in \overline{a}$, that is, $\overline{a} \subseteq \overline{b}$.

We now need to establish the reverse inclusion, $\overline{b} \subseteq \overline{a}$. For this, let $d \in \overline{b}$, that is, $dRb$. As $aRb$, we have, by symmetry, $bRa$. Now $dRb$ and $bRa$ so, by transitivity, $dRa$. Thus, $d \in \overline{a}$ for all $d \in \overline{b}$, that is, $\overline{b} \subseteq \overline{a}$. As we already have the reverse inclusion, $\overline{a} = \overline{b}$.

Now for the 'only if' part. Suppose that $\overline{a} = \overline{b}$. We have to show that $aRb$. But we have seen that $aRa$ is always true so $a \in \overline{a}$. As $\overline{a} = \overline{b}$, we must have $a \in \overline{b}$, that is, $aRb$. ●

- Different equivalence classes do not overlap, for example $\overline{1}$ and $\overline{4}$ have no elements in common. This is the second ingredient for a partition and, as the next theorem shows, holds in general.

## ● Theorem 2

Let $R$ be an equivalence relation on a non-empty set $A$ and let $a, b \in A$. If $\overline{a} \neq \overline{b}$ then $\overline{a} \cap \overline{b} = \emptyset$. ●

PROOF

This is logically equivalent to saying that if $\overline{a} \cap \overline{b} \neq \emptyset$ then $\overline{a} = \overline{b}$. So suppose that $\overline{a} \cap \overline{b} \neq \emptyset$. We can then choose $c \in \overline{a} \cap \overline{b}$. Thus, $c \in \overline{a}$ and $c \in \overline{b}$, that is, $cRa$ and $cRb$. By symmetry, $aRc$. Now $aRc$ and $cRb$ so, by transitivity, $aRb$. By Theorem 1, $\overline{a} = \overline{b}$. ●

## ● Theorem 3

Let $R$ be an equivalence relation on a non-empty set $A$. Then the equivalence classes of $R$ form a partition of $A$. ●

PROOF

We have seen that $A$ is the union of the equivalence classes (because $a \in \overline{a}$) and, by Theorem 2, $\overline{a} \cap \overline{b} = \emptyset$ whenever $\overline{a} \neq \overline{b}$. These are the required properties for a partition. ●

**EXERCISE 4**

Let $R$ be the equivalence relation on $\mathbb{Q}$ given by the rule $aRb \Leftrightarrow a - b \in \mathbb{Z}$. What is the equivalence class of $2\frac{3}{4}$? What is the connection between this and the action of $\mathbb{Z}$ on $\mathbb{Q}$ by the rule $z * q = z + q$?

# 8.4 Equivalence Relations from Group Actions

When a group $G$ acts on a non-empty set $X$ there are resulting equivalence relations and partitions of both $X$ and $G$. Examples of such partitions were discussed at the beginning of Section 8.1 where the group $D_4$ was acting on $X = \{1, 2, 3, 4, 5, 6, 7, 8, 9\}$. The sets in the partition of $X$ in the example were the orbits of the action. It turns out that the orbits of a group action on any set $X$ always form a partition of $X$. To show this, we specify below, for any group action on $X$, an equivalence relation on $X$ for which the equivalence classes are the orbits.

Let $G$ act on $X$. We specify a relation $R$ on $X$ as follows

$aRb$ if and only if there exists $g \in G$ such that $a = g * b$.

Thus, $a$ is related to $b$ when we can get $a$ from $b$ using the action of a group element $g$. We check that $R$ is an equivalence relation.

- Reflexivity: let $a \in X$. Then $a = e * a$ by GA1 so $aRa$.
- Symmetry: let $a, b \in X$ be such that $aRb$, that is, $a = g * b$ for some $g \in G$. We seek an element of $G$ to take us back from $a$ to $b$. The obvious candidate is $g^{-1}$: $g^{-1} * a = g^{-1} * (g * b) = (g^{-1}g) * b$ (by GA2) $= e * b = b$ (by GA1). Thus, $bRa$.
- Transitivity: let $a, b, c \in X$ be such that $aRb$ and $bRc$, that is $a = g * b$ and $b = h * c$ for some $g, h \in G$. Then $gh \in G$ and, by GA2, $(gh) * c = g * (h * c) = g * b = a$ so $aRc$.

Thus, $R$ is an equivalence relation and so its equivalence classes will give a partition of $X$. What are these equivalence classes? Let $a \in X$. Then $\overline{a}$ is the set $\{b \in X : bRa\}$ of all elements of $X$ related to $a$. But, for this relation $R$, this is precisely the set $\{b \in X : b = g * a \text{ for some } g \in G\}$ of all elements of $X$ that can be obtained from $a$ by the group action. This is just the orbit of $a$. Thus, the equivalence classes of $R$ are the orbits of $*$ and so, by Theorem 3, the orbits form a partition of $X$.

## Example 2

Consider the action of the orthogonal group on $\mathbb{R}^2$ by rotation and reflection. The orbit of the origin 0 is just $\{0\}$ but the orbit of any other point $P$ is the circle through $P$ with centre at 0. Thus, this action partitions the plane $\mathbb{R}^2$ into infinitely many concentric circles (one with radius $r$ for each positive real number $r$) together with the singleton $\{0\}$.

Next we look at equivalence relations and partitions of the *group* when a group $G$ acts on a set $X$. In Section 8.1, we discussed such a partition of $D_4$ arising from its action on $\{1, 2, 3, 4, 5, 6, 7, 8, 9\}$ and determined by the effect of each element of the group on the element 1 of $X$. In general, each element of $X$ determines an equivalence relation on $G$ and hence a partition of $G$ into equivalence classes.

Given $x \in X$, we specify a relation $R_x$ on $G$ by the rule

$$gR_xh \text{ if and only if } g * x = h * x.$$

Thus, $g$ is related to $h$ when they have the same effect on $x$.

## Example 3

With $G = D_4$ acting on the set $X = \{1, 2, 3, 4, 5, 6, 7, 8, 9\}$ of nine numbered squares, $r_1 * 1 = 7 = s_1 * 1$ and so $r_1 R_x s_1$ where $x = 1$.

It is clear from its definition that the relation $R_x$ is reflexive, symmetric and transitive and hence an equivalence relation. If $g \in G$ then its equivalence class $\overline{g}$ consists of all those elements $h$ of $G$ such that $h * x = g * x$; that is, those that have the same effect on $x$ as $g$. In the example above, where $x = 1$, $\overline{r_1} = \{r_1, s_1\}$. The partition of $D_4$ into equivalence classes for $R_x$ is the one given in Section 8.1

$$\begin{array}{lllll} \overline{e} = \{e, s_4\} = \overline{s_4} & & \overline{r_1} = \{r_1, s_1\} = \overline{s_1} \\ \overline{r_3} = \{r_3, s_3\} = \overline{s_3} & & \overline{r_2} = \{r_2, s_2\} = \overline{s_2}. \end{array}$$

### Notation

In the above example, and in general, each element $y$ of the orbit of $x$ determines an equivalence class of $R_x$, the set of all elements of $G$ that send $x$ to $y$, that is, the set

$$\{g \in G : g * x = y\}.$$

We shall denote this set by $\mathrm{send}_x(y)$. For example, in the above action of $D_4$, the orbit of 1 is $\{1, 3, 7, 9\}$ and

$$\begin{array}{lllllll} \mathrm{send}_1(1) & = & \{e, s_4\} & & \mathrm{send}_1(3) & = & \{r_3, s_3\} \\ \mathrm{send}_1(7) & = & \{r_1, s_1\} & & \mathrm{send}_1(9) & = & \{r_2, s_2\}. \end{array}$$

Notice that $\mathrm{send}_1(1) = \mathrm{stab}(1)$. Indeed, $\mathrm{send}_x(x)$ is always the stabilizer of $x$.

## EXERCISE 5

Describe the partition of $\mathbb{R}^2$ arising from the action, by matrix multiplication, of the subgroup $H = \left\{\begin{pmatrix} 1 & 0 \\ a & 1 \end{pmatrix} : a \in \mathbb{R}\right\}$ of $GL_2(\mathbb{R})$.

## EXERCISE 6

Consider the action of $D_4$ on the nine numbered squares $\{1, 2, 3, 4, 5, 6, 7, 8, 9\}$. For each element $j$ of the orbit of 4, write down all the elements of $\mathrm{send}_4(j)$.

## EXERCISE 7

Let the group $G$ act on the non-empty set $X$, let $x \in X$ and let $H = \mathrm{stab}(x)$ be the stabilizer of $x$. Let $a \in G$ and let $y = a * x$. (Thus, $y \in \mathrm{orb}(x)$.)

(i) Let $b \in \mathrm{send}_x(y)$, that is, $b * x = y = a * x$. Show that $a^{-1}b \in H$ and hence that $b = ah$ for some $h \in H$.

(ii) Let $h \in H$. Show that $ah \in \mathrm{send}_x(y)$.

(This exercise shows that $\mathrm{send}_x(y)$ is the set of all elements of $G$ of the form $ah$, $h \in H$. This set is denoted $aH$. We will meet such sets again in Chapter 10.)

## 8.5 Modular Arithmetic

Let $n > 1$ be an integer. The set of all equivalence classes of equivalence modulo $n$ will be denoted $\mathbb{Z}_n$. For example, $Z_5 = \{\overline{0}, \overline{1}, \overline{2}, \overline{3}, \overline{4}\}$. We aim to define binary operations addition and multiplication modulo $n$ on $\mathbb{Z}_n$ by the rules

$$\overline{a} + \overline{b} = \overline{a + b}; \quad \overline{a} \times \overline{b} = \overline{a \times b}.$$

However, there could be a problem of ambiguity here arising from the fact that different values of $a$ give the same $\overline{a}$ and similarly for $b$. For example, when $n = 5$, $\overline{2} = \overline{7}$ and

$\overline{13} = \overline{3}$ so $\overline{2} + \overline{13}$ must equal $\overline{7} + \overline{3}$ and $\overline{2} \times \overline{13}$ must equal $\overline{7} \times \overline{3}$ if these operations are to be unambiguous. In our numerical example

$$\overline{2+13} = \overline{15} = \overline{0},\ \overline{7+3} = \overline{10} = \overline{0},\ \overline{2 \times 13} = \overline{26} = \overline{1} \text{ and } \overline{7 \times 3} = \overline{21} = \overline{1}$$

so the problem does not arise here. However, we need to check that it never arises. To do this, let $a, b, c, d \in \mathbb{Z}$ be such that $\overline{a} = \overline{c}$ and $\overline{b} = \overline{d}$. We have to check that $\overline{a+b} = \overline{c+d}$ and $\overline{ab} = \overline{cd}$. By Theorem 1, $c \equiv a \bmod n$ and $d \equiv b \bmod n$ so there are integers $p, q$ such that

$$c = a + np \text{ and } d = b + nq.$$

Then

$$(c+d) = (a+b) + n(p+q)$$

so $(c+d) \equiv (a+b) \bmod n$ and, by Theorem 1, $\overline{c+d} = \overline{a+b}$. Thus, addition modulo $n$ is well-defined.

For multiplication

$$cd = (a+np)(b+nq) = ab + n(pb + aq + npq)$$

so $(cd) \equiv (ab) \bmod n$ and, by Theorem 1, $\overline{cd} = \overline{ab}$. Thus multiplication modulo $n$ is also well-defined.

If $m$ is any integer then, dividing $m$ by $n$, we can write $m = qn + r$ where $q, r \in \mathbb{Z}$ and the remainder $r$ satisfies $0 \leqslant r < n$. Then $m \equiv r \bmod n$ and, by Theorem 1, $\overline{m} = \overline{r}$. So every equivalence class appears in the list $\overline{0}, \overline{1}, \overline{2}, \ldots, \overline{n-1}$. Moreover, these $n$ classes are all different because if $0 \leqslant r < s < n$ then $0 < s - r < n$ so it is impossible for $n$ to divide $s - r$. Theorem 1 then tells us that $\overline{r}$ cannot equal $\overline{s}$. Thus, as we have seen for $n = 5$, there are precisely $n$ different equivalence classes in $\mathbb{Z}_n$.

Tables for addition and multiplication modulo 5 in $\mathbb{Z}_5 = \{\overline{0}, \overline{1}, \overline{2}, \overline{3}, \overline{4}\}$ are shown in Table 8.1 where we have written all equivalence classes in the form $\overline{r}$, $0 \leqslant r < 5$, for example $\overline{3} \times \overline{4} = \overline{12} = \overline{2}$.

| +mod5 | $\overline{0}$ | $\overline{1}$ | $\overline{2}$ | $\overline{3}$ | $\overline{4}$ |
|---|---|---|---|---|---|
| $\overline{0}$ | $\overline{0}$ | $\overline{1}$ | $\overline{2}$ | $\overline{3}$ | $\overline{4}$ |
| $\overline{1}$ | $\overline{1}$ | $\overline{2}$ | $\overline{3}$ | $\overline{4}$ | $\overline{0}$ |
| $\overline{2}$ | $\overline{2}$ | $\overline{3}$ | $\overline{4}$ | $\overline{0}$ | $\overline{1}$ |
| $\overline{3}$ | $\overline{3}$ | $\overline{4}$ | $\overline{0}$ | $\overline{1}$ | $\overline{2}$ |
| $\overline{4}$ | $\overline{4}$ | $\overline{0}$ | $\overline{1}$ | $\overline{2}$ | $\overline{3}$ |

| ×mod5 | $\overline{0}$ | $\overline{1}$ | $\overline{2}$ | $\overline{3}$ | $\overline{4}$ |
|---|---|---|---|---|---|
| $\overline{0}$ | $\overline{0}$ | $\overline{0}$ | $\overline{0}$ | $\overline{0}$ | $\overline{0}$ |
| $\overline{1}$ | $\overline{0}$ | $\overline{1}$ | $\overline{2}$ | $\overline{3}$ | $\overline{4}$ |
| $\overline{2}$ | $\overline{0}$ | $\overline{2}$ | $\overline{4}$ | $\overline{1}$ | $\overline{3}$ |
| $\overline{3}$ | $\overline{0}$ | $\overline{3}$ | $\overline{1}$ | $\overline{4}$ | $\overline{2}$ |
| $\overline{4}$ | $\overline{0}$ | $\overline{4}$ | $\overline{3}$ | $\overline{2}$ | $\overline{1}$ |

**Table 8.1** Arithmetic modulo 5.

The left-hand table looks like the Cayley table of a group but, because of all those $\overline{0}$s, the right-hand one does not. Let us check on the group axioms for addition and multiplication modulo $n$. It is clear from the definitions that both are closed, that is G1 holds. Both are associative because of the associativity of the corresponding binary operation in $\mathbb{Z}$. For example, for addition modulo $n$

$$(\overline{a} + \overline{b}) + \overline{c} \quad = \quad \overline{a+b} + \overline{c}$$

$$
\begin{aligned}
&= \overline{(a+b)+c} \\
&= \overline{a+(b+c)} \\
&= \overline{a} + \overline{b+c} \\
&= \overline{a} + (\overline{b} + \overline{c}).
\end{aligned}
$$

Thus, G2 holds for addition modulo $n$ and, replacing $+$ by $\times$ throughout the above calculation, it also holds for multiplication modulo $n$.

Clearly $\overline{0}$ is neutral for addition modulo $n$ and $\overline{1}$ is neutral for multiplication modulo $n$. Thus, G3 holds for both.

This brings us to inverses. For addition modulo $n$, there is no problem; each element $\overline{a}$ of $\mathbb{Z}_n$ has inverse $\overline{-a} = \overline{n-a}$. For example, with $n = 5$, $\overline{2} + \overline{3} = \overline{0}$ so $\overline{2}$ has inverse $\overline{3}$. Thus, $\mathbb{Z}_n$ satisfies all four group axioms and is a group under addition modulo $n$. It is clearly abelian and, in fact, is cyclic. As it is in additive notation, the cyclic subgroup generated by $\overline{a}$ consists of all elements of the form $\underbrace{\overline{a} + \overline{a} + \ldots + \overline{a}}$. In particular, the cyclic subgroup generated by $\overline{1}$ is the whole group $\mathbb{Z}_n$.

In a sense to be formalized in the next chapter, $\mathbb{Z}_n$ is the prototype for all cyclic groups $G = \langle g \rangle = \{e, g, g^2, \ldots, g^{n-1}\}$ of order $n$. When any two elements $g^i$ and $g^j$ are multiplied together

$$g^i g^j = g^{i+j} = g^r$$

where $r$ is the remainder on division of $i + j$ by $n$. We are effectively performing addition modulo $n$ on the indices. We can rephrase parts of Theorem 3 of Chapter 6 in terms of equivalence modulo $n$ as follows.

- For $m \in \mathbb{Z}$, $g^m = e$ if and only if $m \equiv 0 \bmod n$.
- For $i, j \in \mathbb{Z}$, $g^i = g^j$ if and only if $i = j \bmod n$ if and only if $\overline{i} = \overline{j}$.

For multiplication modulo $n$, where the neutral element is $\overline{1}$, the situation is different because $\overline{0} \times \overline{a} = \overline{0} \neq \overline{1}$ for all $\overline{a} \in \mathbb{Z}_n$ so $\overline{0}$ has no inverse. When we met the corresponding problem for $\mathbb{R}$ under multiplication, we omitted 0 to obtain the group $\mathbb{R}\backslash\{0\}$. So let us try the corresponding thing for $\mathbb{Z}_n$, that is, consider $\mathbb{Z}_n\backslash\{\overline{0}\}$. Associativity holds here because it holds in the larger set $\mathbb{Z}_n$ and there is a neutral element $\overline{1}$. But do we have closure and do we have inverses? For $n = 5$ the answer to both questions is yes as can be seen from Table 8.2.

| ×mod5 | $\overline{1}$ | $\overline{2}$ | $\overline{3}$ | $\overline{4}$ |
|---|---|---|---|---|
| $\overline{1}$ | $\overline{1}$ | $\overline{2}$ | $\overline{3}$ | $\overline{4}$ |
| $\overline{2}$ | $\overline{2}$ | $\overline{4}$ | $\overline{1}$ | $\overline{3}$ |
| $\overline{3}$ | $\overline{3}$ | $\overline{1}$ | $\overline{4}$ | $\overline{2}$ |
| $\overline{4}$ | $\overline{4}$ | $\overline{3}$ | $\overline{2}$ | $\overline{1}$ |

| ×mod6 | $\overline{1}$ | $\overline{2}$ | $\overline{3}$ | $\overline{4}$ | $\overline{5}$ |
|---|---|---|---|---|---|
| $\overline{1}$ | $\overline{1}$ | $\overline{2}$ | $\overline{3}$ | $\overline{4}$ | $\overline{5}$ |
| $\overline{2}$ | $\overline{2}$ | $\overline{4}$ | $\overline{0}$ | $\overline{2}$ | $\overline{4}$ |
| $\overline{3}$ | $\overline{3}$ | $\overline{0}$ | $\overline{3}$ | $\overline{0}$ | $\overline{3}$ |
| $\overline{4}$ | $\overline{4}$ | $\overline{2}$ | $\overline{0}$ | $\overline{4}$ | $\overline{2}$ |
| $\overline{5}$ | $\overline{5}$ | $\overline{4}$ | $\overline{3}$ | $\overline{2}$ | $\overline{1}$ |

**Table 8.2** Multiplication modulo 5 and 6.

So $\mathbb{Z}_5\backslash\{\overline{0}\}$ is a group under multiplication 5. But $\mathbb{Z}_6\backslash\{\overline{0}\}$ is not closed because of the appearances of $\overline{0}$. Moreover, the elements $\overline{2}$, $\overline{3}$ and $\overline{4}$ have no inverses. We have encountered the latter problem before with $\mathbb{Z}$, where only 1 and $-1$ have inverses for multiplication, and with $2 \times 2$ real matrices, where not all non-zero matrices are invertible. In both cases we obtained a group by taking only the elements which do have inverses. This suggests that to get a group under multiplication modulo $n$ we should take the set

$$U(\mathbb{Z}_n) = \{\overline{a} \in \mathbb{Z}_n : \overline{a}\,\overline{b} = \overline{1} \text{ for some } \overline{b} \in \mathbb{Z}_n\}.$$

For example, $U(\mathbb{Z}_5) = \{\overline{1}, \overline{2}, \overline{3}, \overline{4}\}$, which we have already seen to be a group, and $U(\mathbb{Z}_6) = \{\overline{1}, \overline{5}\}$.

| $\times$mod6 | $\overline{1}$ | $\overline{5}$ |
|---|---|---|
| $\overline{1}$ | $\overline{1}$ | $\overline{5}$ |
| $\overline{5}$ | $\overline{5}$ | $\overline{1}$ |

**Table 8.3** Group under multiplication modulo 6.

The multiplication table for the latter is shown in Table 8.3 from which it is clear that it is a cyclic group of order 2.

We now check that $U(\mathbb{Z}_n)$ is always a group. If $\overline{a} \in U(\mathbb{Z}_n)$, with inverse $\overline{b}$, and $\overline{c} \in U(\mathbb{Z}_n)$, with inverse $\overline{d}$ then $\overline{a}\,\overline{c} \in U(\mathbb{Z}_n)$ with inverse $\overline{bd}$. Thus, we have closure of $U(\mathbb{Z}_n)$. We have seen that multiplication modulo $n$ is associative. A neutral element is $\overline{1}$ and if $\overline{a} \in U(\mathbb{Z}_n)$ with $\overline{a}\overline{b} = \overline{1}$ then $\overline{b} \in U(\mathbb{Z}_n)$ and is an inverse for $\overline{a}$. We have shown that $U(\mathbb{Z}_n)$ is indeed a group for all $n$. It is clearly abelian.

We have seen that $U(\mathbb{Z}_5) = \{\overline{1}, \overline{2}, \overline{3}, \overline{4}\} = \mathbb{Z}_5 \backslash \{\overline{0}\}$ whereas $U(\mathbb{Z}_6) = \{\overline{1}, \overline{5}\} \neq \mathbb{Z}_6 \backslash \{\overline{0}\}$. The crucial distinction between 5 and 6 is that 5 is prime but 6 is not. The fact that $|U(\mathbb{Z}_5)| = 4$ is a special case of the next theorem.

## ● Theorem 4

Let $p$ be a prime number. Then every element $\overline{a}$ of $\mathbb{Z}_p \backslash \{\overline{0}\}$ has an inverse and so $U(\mathbb{Z}_p)$ has order $p - 1$. ●

PROOF

Let $\overline{a}, \overline{b} \in \mathbb{Z}_p \backslash \{\overline{0}\}$. Thus, $p$ divides neither $a$ nor $b$. Because $p$ is prime, it does not divide $ab$ and so $\overline{a}\,\overline{b} \neq \overline{0}$. Taking $b = 1, 2, \ldots, p - 1$, we obtain $p - 1$ elements of $\mathbb{Z}_p \backslash \{\overline{0}\}$. Moreover, they are distinct. To see this, let $1 \leqslant b < c \leqslant p - 1$. Then $1 \leqslant c - b \leqslant p - 1$. Thus, $p$ divides neither $a$ nor $c - b$ and so $\overline{a}\,(\overline{c - b}) \neq \overline{0}$. Thus, $ac - ab = a(c - b)$ is not a multiple of $p$ and hence $\overline{a}\,\overline{c} \neq \overline{a}\,\overline{b}$. Thus, the $p - 1$ elements $\overline{a}\,\overline{b}$ are different and must be all the elements of $\mathbb{Z}_p \backslash \{\overline{0}\}$. One of them must then be $\overline{1}$ and so $\overline{a}$ has an inverse. ●

Some readers will be familiar with Euclid's Algorithm which finds the highest common factor of two positive integers $p$ and $a$ and expresses it in the form $ps + at$ for some integers $s$ and $t$. With $p$ prime and $1 \leqslant a \leqslant p - 1$ as in the proof of Theorem 4, the h.c.f. must be 1 and $\overline{t}$ is then an inverse for $\overline{a}$. This gives an alternative proof of the theorem.

**EXERCISE 8**

For each of $n = 3, 4, 8$ decide which elements of $\mathbb{Z}_n$ are in $U(\mathbb{Z}_n)$.

## 8.6 Further Exercises on Chapter 8

**EXERCISE 9**

We define a relation $R$, called equivalence modulo $2\pi$, on the set $\mathbb{R}$ of all real numbers by the rule

$$\theta R \phi \Leftrightarrow \theta - \phi = 2n\pi \text{ for some } n \in \mathbb{Z}.$$

Thus, $aRb$ precisely when $\cos\theta = \cos\phi$ and $\sin\theta = \sin\phi$. Show that $R$ is an equivalence relation on $\mathbb{R}$.

**EXERCISE 10**

Let $R$ be the relation on $\mathbb{Z}$ given by the rule

$$aRb \Leftrightarrow a \equiv 5 \bmod b.$$

Show that $R$ is not reflexive, not symmetric and not transitive.

**EXERCISE 11**

Describe the partition of $\mathbb{R}^2$ arising from the orbits of the action, by matrix multiplication, of each of the following subgroups $H$ of $GL_2(\mathbb{R})$.

(i) $H = \left\{ \begin{pmatrix} \cos\theta & -2\sin\theta \\ \frac{1}{2}\sin\theta & \cos\theta \end{pmatrix} : \theta \in \mathbb{R} \right\}$;

(ii) $H = \left\{ \begin{pmatrix} a & b \\ b & a \end{pmatrix} : a, b \in \mathbb{R} \text{ and } a^2 - b^2 = 1 \right\}$.

**EXERCISE 12**

Let $P = \begin{pmatrix} 1 \\ 0 \end{pmatrix}$, $Q = \begin{pmatrix} 1 \\ 1 \end{pmatrix} \in \mathbb{R}^2$. Show that, for the action of the group $GL_2(\mathbb{R})$ of all $2 \times 2$ invertible real matrices on $\mathbb{R}^2$ by matrix multiplication, $\text{send}_P(Q)$ is the set of all $2 \times 2$ matrices of the form $\begin{pmatrix} 1 & b \\ 1 & d \end{pmatrix}$ where $b, d \in \mathbb{R}$ and $b \neq d$.

**EXERCISE 13**

Let $L$ be the line $y = x$ and let $L'$ be the line $y = -x$. For the action of the orthogonal group $O_2$ on the set of straight lines through the origin, find $\text{send}_L(L')$.

## EXERCISE 14

Show that $U(\mathbb{Z}_7)$ is a cyclic group generated by $\overline{3}$.

## EXERCISE 15

Show that $U(\mathbb{Z}_8)$ and $U(\mathbb{Z}_{10})$ both have order 4. Compute their Cayley tables and show that one of them is cyclic but the other is not.

## EXERCISE 16

Investigate the group $U(\mathbb{Z}_n)$ for various values of $n$. Suggest rules for when $\overline{a} \in U(\mathbb{Z}_n)$ and for when every element of $\mathbb{Z}_n$, except $\overline{0}$, is in $U(\mathbb{Z}_n)$.

# 9 • Homomorphisms and Isomorphisms

We have occasionally commented that two groups are in some sense the same. One purpose of this chapter is to formalize such comments. We shall do this by looking at functions from one group to another which preserve the structure of the groups.

## 9.1 Comparing $D_3$ and $S_3$

Below, we reproduce the Cayley tables for the dihedral group $D_3$ of symmetries of the equilateral triangle and the group $S_3$ of permutations of $\{1, 2, 3\}$. In the table for $S_3$, we have abbreviated the permutations as follows

$$(1\ 2\ 3) = \rho_1,\ (1\ 3\ 2) = \rho_2,\ (1\ 2) = \sigma_1,\ (1\ 3) = \sigma_2,\ (2\ 3) = \sigma_3.$$

In the table for $D_3$

$$r_1 = \text{rot}_{\frac{2\pi}{3}},\ r_2 = \text{rot}_{\frac{4\pi}{3}},\ s_1 = \text{ref}_0,\ s_2 = \text{ref}_{\frac{2\pi}{3}},\ s_3 = \text{ref}_{\frac{4\pi}{3}}.$$

These tables appear the same except for the alphabet used and it is clear that, in some sense, the two groups are the same. One common feature is that they have the same

| **D₃** | $e$ | $r_1$ | $r_2$ | $s_1$ | $s_2$ | $s_3$ |
|---|---|---|---|---|---|---|
| $e$ | $e$ | $r_1$ | $r_2$ | $s_1$ | $s_2$ | $s_3$ |
| $r_1$ | $r_1$ | $r_2$ | $e$ | $s_2$ | $s_3$ | $s_1$ |
| $r_2$ | $r_2$ | $e$ | $r_1$ | $s_3$ | $s_1$ | $s_2$ |
| $s_1$ | $s_1$ | $s_3$ | $s_2$ | $e$ | $r_2$ | $r_1$ |
| $s_2$ | $s_2$ | $s_1$ | $s_3$ | $r_1$ | $e$ | $r_2$ |
| $s_3$ | $s_3$ | $s_2$ | $s_1$ | $r_2$ | $r_1$ | $e$ |

| **S₃** | $e$ | $\rho_1$ | $\rho_2$ | $\sigma_1$ | $\sigma_2$ | $\sigma_3$ |
|---|---|---|---|---|---|---|
| $e$ | $e$ | $\rho_1$ | $\rho_2$ | $\sigma_1$ | $\sigma_2$ | $\sigma_3$ |
| $\rho_1$ | $\rho_1$ | $\rho_2$ | $e$ | $\sigma_2$ | $\sigma_3$ | $\sigma_1$ |
| $\rho_2$ | $\rho_2$ | $e$ | $\rho_1$ | $\sigma_3$ | $\sigma_1$ | $\sigma_2$ |
| $\sigma_1$ | $\sigma_1$ | $\sigma_3$ | $\sigma_2$ | $e$ | $\rho_2$ | $\rho_1$ |
| $\sigma_2$ | $\sigma_2$ | $\sigma_1$ | $\sigma_3$ | $\rho_1$ | $e$ | $\rho_2$ |
| $\sigma_3$ | $\sigma_3$ | $\sigma_2$ | $\sigma_1$ | $\rho_2$ | $\rho_1$ | $e$ |

**Table 9.1** Cayley tables for $D_3$ and $S_3$.

number of elements. Given any two finite groups of the same order, we can construct a bijective function from one to the other, pairing off the elements. However, such a bijective function need have nothing to do with the binary operations in the two groups. Two groups of the same order can behave differently. Can you think of a difference between the group of rotations of a hexagon and the group $D_3$?

For the two groups $D_3$ and $S_3$ there is a bijective function from $D_3$ to $S_3$ which behaves well with respect to the binary operations. Let $f : D_3 \to S_3$ be the function given by

$$\begin{array}{lcllcl} e & \mapsto & \text{id} & \text{ref}_0 & \mapsto & (1\,2) \\ \text{rot}_{\frac{2\pi}{3}} & \mapsto & (1\,2\,3) & \text{ref}_{\frac{2\pi}{3}} & \mapsto & (1\,3) \\ \text{rot}_{\frac{4\pi}{3}} & \mapsto & (1\,3\,2) & \text{ref}_{\frac{4\pi}{3}} & \mapsto & (2\,3). \end{array}$$

Thus, for each $x \in D_3$, $f(x)$ is the corresponding permutation of the vertices of the equilateral triangle numbered as shown in Fig 9.1. Choose any pair of elements

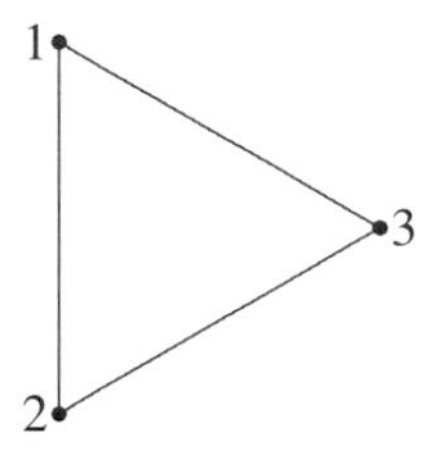


**Fig 9.1** Vertices of the equilateral triangle.

$x, y \in D_3$. You should find that $f(x)f(y) = f(xy)$, whatever pair you choose. In other words, if you combine $x$ and $y$ in $D_3$ and then apply $f$ you get the same result as when you apply $f$ to each and then combine the resulting elements $f(x)$ and $f(y)$ of $S_3$. Not only does $f$ pair off the elements of the two groups but the whole of the Cayley table for $S_3$ can be obtained from that of $D_3$ by applying $f$. Everywhere where $r_2$ appears in the table for $D_3$, $f(r_2) = \rho_2$ appears in the table for $S_3$.

We now have two features of our function $f : D_3 \to S_3$. There is its bijectivity, giving a correspondence between the *elements* of the two groups and there is the property $f(xy) = f(x)f(y)$ giving a correspondence between the *operations* in the two groups. This leads us to the following definitions.

Let $G$ and $H$ be groups and $f$ be a function $f : G \to H$. Then $f$ is said to be a **homomorphism** if $f(g_1g_2) = f(g_1)f(g_2)$ for all $g_1, g_2 \in G$. A bijective

| $G$ | $g_2$ |
|---|---|
| $g_1$ | $g_1g_2$ |

| $H$ | $f(g_2)$ |
|---|---|
| $f(g_1)$ | $f(g_1)f(g_2)$ $= f(g_1g_2)$ |

**Table 9.2** Homomorphism.

homomorphism is called an **isomorphism**.

We say that two groups $G$, $H$ are **isomorphic** and write $G \cong H$ if there is an isomorphism $f : G \to H$. The words isomorphism and homomorphism are derived from the Greek

iso (same) homo (like) morphism (shape).

For example, the dihedral group $D_3$ is isomorphic to the symmetric group $S_3$. That our function $f : D_3 \to S_3$ is a homomorphism is clear from inspection of the Cayley tables. We shall see in Section 9.3 that it is one of a general class of homomorphisms arising from group actions.

There are infinitely many equilateral triangles with centres at the origin. Their groups of symmetries are all isomorphic to $D_3$. For example, if the $y$-axis rather than the $x$-axis is a line of symmetry of the triangle, then the group of symmetries is
$\{e, \mathrm{rot}_{\frac{2\pi}{3}}, \mathrm{rot}_{\frac{4\pi}{3}}, \mathrm{ref}_{\frac{\pi}{3}}, \mathrm{ref}_{\pi}, \mathrm{ref}_{\frac{5\pi}{3}}\}$ which is isomorphic to $D_3$ with

$$e \mapsto e,\ \mathrm{rot}_{\frac{2\pi}{3}} \mapsto \mathrm{rot}_{\frac{2\pi}{3}},\ \mathrm{rot}_{\frac{4\pi}{3}} \mapsto \mathrm{rot}_{\frac{4\pi}{3}},\ \mathrm{ref}_{\frac{\pi}{3}} \mapsto \mathrm{ref}_0,\ \mathrm{ref}_{\pi} \mapsto \mathrm{ref}_{\frac{2\pi}{3}},\ \text{and } \mathrm{ref}_{\frac{5\pi}{3}} \mapsto \mathrm{ref}_{\frac{4\pi}{3}}.$$

## Klein's 4-group

We have seen several groups $G = \{e, a, b, c\}$ of order 4 with Cayley tables of the form below. Probably the most familiar example is the group of symmetries of the rectangle

| G | $e$ | $a$ | $b$ | $c$ |
|---|---|---|---|---|
| $e$ | $e$ | $a$ | $b$ | $c$ |
| $a$ | $a$ | $e$ | $c$ | $b$ |
| $b$ | $b$ | $c$ | $e$ | $a$ |
| $c$ | $c$ | $b$ | $a$ | $e$ |

**Table 9.3** Klein's 4-group.

where $e = \mathrm{rot}_0$, $a = \mathrm{rot}_\pi$, $b = \mathrm{ref}_0$ and $c = \mathrm{ref}_\pi$. Others have appeared in exercises. The stabilizer of the line $y = x$ for the action of the orthogonal group $O_2$ on straight lines through the origin has this form with

$$a = \mathrm{rot}_0,\ b = \mathrm{ref}_{\frac{\pi}{2}} \text{ and } c = \mathrm{ref}_{\frac{3\pi}{2}}.$$

Another example is the stabilizer of the polynomial $x_1x_2 + x_3 + x_4$ for the action of $S_4$, where

$$a = (3\ 4),\ b = (1\ 2) \text{ and } c = (1\ 2)(3\ 4).$$

A further example is the group $U(\mathbb{Z}_8)$ under multiplication modulo 8 where

$$a = \overline{3},\ b = \overline{5} \text{ and } c = \overline{7}.$$

Clearly any two such groups are isomorphic. As they are essentially the same, we use the same name, **Klein's 4-group** after **F. Klein**, for any of them. This is the first of several instances where we give the same name to isomorphic groups even though they have different elements. When two groups are isomorphic, we are unable to distinguish between them group theoretically.

The letter $V$ is often used to denote Klein's 4-group (from the German *vier*=four). Although $V$ is not cyclic, it is isomorphic to a direct product of cyclic groups. To see this, let $G =< g >$ be a cyclic group of order 2 and consider the direct product $G \times G$. This has four elements, namely $(e, e)$, $(g, e)$, $(e, g)$ and $(g, g)$ and fits the above description of Klein's 4-group with $a = (g, e)$, $b = (e, g)$ and $c = (g, g)$.

In defining homomorphism and isomorphism, we followed our convention of supressing the binary operation and using juxtaposition. If the binary operations in $G$ and $H$ are $\odot$ and $\oplus$ respectively, then the defining property for a homomorphism $f : G \to H$ becomes

$$f(g_1 \odot g_2) = f(g_1) \oplus f(g_2).$$

The most likely situation where we might need this is where one group is in additive notation, as in the next example.

## Example 1

Let $G$ be the group $\mathbb{R}$ under addition and let $H = \left\{ \begin{pmatrix} 1 & a \\ 0 & 1 \end{pmatrix} : a \in \mathbb{R} \right\}$, that is the group of $2 \times 2$ matrices representing shears parallel to the $x$-axis. Let $f : G \to H$ be the function such that

$$f : a \mapsto \begin{pmatrix} 1 & a \\ 0 & 1 \end{pmatrix}.$$

Then $f$ is clearly bijective. Here $G$, but not $H$, is in additive notation so to show that $f$ is an isomorphism, we need to check that $f(a + b) = f(a)f(b)$ for all $a, b \in G$. This is a routine exercise in matrix multiplication. Thus, $G \cong H$.

So far we have concentrated on *iso*morphisms. Some non-bijective homomorphisms have already appeared in the book and we now look at these.

## Example 2

Let $G = GL_2(\mathbb{R})$, the group of $2 \times 2$ invertible matrices and let $H = (\mathbb{R}\backslash\{0\}, \times)$, the group of non-zero real numbers under multiplication. Let $f : G \to H$ be the function with $f(A) = \det A$. From Theorem 1(ii) of Chapter 3, we know that $f(AB) = \det(AB) = \det A \det B = f(A)f(B)$. Thus, $f$ is a homomorphism.

## Example 3

Let $n \geqslant 2$, let $G$ be the symmetric group $S_n$ and let $H$ be the group of non-zero reals under multiplication. So far we have mostly used the letters $f$ and $g$ for permutations. To avoid possible confusion with the use of $f$ for homomorphisms, we shall sometimes use Greek letters for permutations. Consider the function $f : G \to H$ where $f(\alpha) = \text{sgn}(\alpha)$ for all $\alpha \in S_n$. The rule $\text{sgn}(\alpha\beta) = \text{sgn}\,\alpha\ \text{sgn}\,\beta$, which is established on page 76, says that $f$ is a homomorphism.

## EXERCISE 1

Let $f$ be the determinant homomorphism from $G = GL_2(\mathbb{R})$ to $(\mathbb{R}\backslash\{0\}, \times)$. Why is $f$ not injective? Is it surjective?

**EXERCISE 2**

Let $f$ be the sign homomorphism from $S_n$ to $(\mathbb{R}\backslash\{0\}, \times)$. Why is $f$ not surjective? Show that if $n > 2$ then $f$ is not injective. Is it injective when $n = 2$?

**EXERCISE 3**

Let $G = (\mathbb{R}, +)$ and let $H = \mathbb{R}^+$ be the group of positive real numbers under multiplication. Show that the function $f : G \to H$ given by $f : x \mapsto \mathrm{e}^x$ is an isomorphism. What is its inverse? Check that its inverse is an isomorphism from $H$ to $G$.

## 9.2 Properties of Homomorphisms

In this section we establish some basic properties of homomorphisms and isomorphisms. The first result justifies the inherent symmetry in the phrase '$G$ and $H$ are isomorphic'. An example of this with $G = (\mathbb{R}, +)$, $H = \mathbb{R}^+$ and $f : x \mapsto \mathrm{e}^x$ appeared in an exercise above.

● *Theorem 1*

Let $f : G \to H$ be an isomorphism. Then $f^{-1} : H \to G$ is an isomorphism. ●

PROOF

Since $f$ is a bijection, $f^{-1}$ exists and is a bijection. Let $h, k \in H$. Then there are $p, q \in G$ with $f(p) = h$ and $f(q) = k$. Now
$f^{-1}(hk) = f^{-1}(f(p)f(q)) = f^{-1}(f(pq)) = pq = f^{-1}(h)f^{-1}(k)$. Hence $f^{-1}$ is an isomorphism. ●

If $G$ is any group then the identity function $\mathrm{id} : G \to G$ is clearly a homomorphism. It is certainly bijective so it is an isomorphism and $G \cong G$. If $H$ is a subgroup of $G$ then there is a homomorphism $f : H \to G$ given by $f : h \mapsto h$. This homomorphism, which is injective but, unless $H = G$, not surjective, is sometimes called the **inclusion** of $H$ in $G$.

● *Theorem 2*

Let $G$ and $H$ be groups and let $f : G \to H$ be a homomorphism.

(i) $f(e_G) = e_H$.

(ii) $f(g^{-1}) = (f(g))^{-1}$ for all $g \in G$.

(iii) If $G$ is abelian and $f$ is surjective (in particular if $f$ is an isomorphism) then $H$ is abelian. ●

PROOF

(i) $f(e_G)f(e_G) = f(e_Ge_G) = f(e_G) = f(e_G)e_H$. The result follows on cancelling $f(e_G)$.

(ii) $f(g)f(g^{-1}) = f(gg^{-1}) = f(e_G) = e_H$ and similarly $f(g^{-1})f(g) = e_H$ so $f(g^{-1})$ is the inverse in $H$ of $f(g)$.

(iii) Suppose that $G$ is abelian and that $f$ is surjective. Let $h, k \in H$. Then there are $a, b \in G$ with $f(a) = h$ and $f(b) = k$. Hence, since $G$ is abelian

$$hk = f(a)f(b) = f(ab) = f(ba) = f(b)f(a) = kh.$$

Thus, $hk = kh$ for all $h, k \in H$, that is, $H$ is abelian. ●

Let $f : G \to H$ be a homomorphism and let $g \in G$. Then, setting $g_1 = g_2 = g$ in the definition of homomorphism, $f(g^2) = (f(g))^2$. This extends to the rule

$$f(g^n) = (f(g))^n$$

for all positive integers $n$. If you are familiar with proof by induction, this is easily proved by that method. By Theorem 2(i) and (ii), this formula also holds when $n = 0$ and when $n$ is negative.

### ● Theorem 3

Let $G$ and $H$ be groups with an isomorphism $f : G \to H$. If $g \in G$ has finite order $n$ in $G$ then $f(g)$ has order $n$ in $H$. ●

PROOF

The formula above tells us that $(f(g))^n = f(g^n) = f(e_G) = e_H$. Hence $f(g)$ has finite order, $m$ say, and $n \geqslant m$. Applying the same argument with $f^{-1}$ replacing $f$ and $f(g)$ replacing $g$ gives $m \geqslant n$. Hence, $m = n$ as required. ●

In Section 9.1 we asked you to think of a difference between two groups of order six, the dihedral group $D_3$ and the group of rotations of a hexagon. One difference is that the latter is abelian and the former is not. By Theorem 2, it follows that the two are not isomorphic. Alternatively, the rotation group has an element $\mathrm{rot}_{\frac{\pi}{3}}$ of order 6 and $D_3$ has no such element so Theorem 3 can be used to demonstrate that they are not isomorphic.

**EXERCISE 4**

Let $G, H, K$ be groups such that $G \cong H$ and $H \cong K$ with isomorphisms $f : G \to H$ and $g : H \to K$. By considering the composite $gf$, show that $G \cong K$.

**EXERCISE 5**

Let $G = \{e, \mathrm{rot}_{\frac{\pi}{2}}, \mathrm{rot}_{\pi}, \mathrm{rot}_{\frac{3\pi}{2}}\}$ be the rotation group of the square. Show that $G$ is not isomorphic to Klein's 4-group.

## 9.3 Homomorphisms Arising from Group Actions

In Section 9.1, we saw that the function $f : D_3 \to S_3$, where $f(g)$ is the permutation of the vertices of the equilateral triangle determined by $g$, is an isomorphism. We can try the same thing with $D_4$ and $S_4$. So now let $f : D_4 \to S_4$ be the function such that, for $g \in D_4$, $f(g)$ is the permutation of the vertices of the square corresponding to $g$. Thus

$$f : e \mapsto \text{id}, \quad f : r_1 \mapsto (1\,2\,3\,4), \quad f : r_2 \mapsto (1\,3)(2\,4), \quad f : r_3 \mapsto (1\,4\,3\,2),$$
$$f : s_1 \mapsto (1\,4)(2\,3), \quad f : s_2 \mapsto (2\,4), \quad f : s_3 \mapsto (1\,2)(3\,4), \quad f : s_4 \mapsto (1\,3).$$

Now $|S_4| = 24$ and $|D_4| = 8$ so $f$ cannot be bijective. Hence, it cannot be an isomorphism. But is it a homomorphism? It would be tedious to check that $f(g_1g_2) = f(g_1)f(g_2)$ for all 64 choices of $g_1$ and $g_2$. Fortunately, $f$ is one of a general class of functions arising from group actions for which the homomorphism property is readily checked.

Recall from page 73 that when a group $G$ acts on a set $X$, to each element $g$ of $G$ there is associated a bijective function $f_g : X \to X$ given by the rule $f_g(x) = g * x$ for all $x \in X$. Moreover we saw that $f_g f_h = f_{gh}$ for all $g, h \in G$. Now each of the functions $f_g$ is in $S_X$ so we have a function $f : G \to S_X$ given by the rule $f(g) = f_g$. As $f_g f_h = f_{gh}$ for all $g, h \in G$, this function $f : G \to S_X$ is a homomorphism. The isomorphism $f$ which we had between $D_3$ and $S_3$ arises in this way from the action of $D_3$ on the vertices of the equilateral triangle. When $G$ is $D_4$ acting on the vertices of the square, this homomorphism $f$ is the function displayed above.

As we have commented above, $D_4$ and $S_4$ cannot be isomorphic as their orders are different. However, $D_4$ is isomorphic to a subgroup of $S_4$. To see this, we extend the idea of image, which we met in the context of symmetries in Chapter 5, to arbitrary functions, including homomorphisms.

Let $f : A \to B$ be a function and let $C$ be a subset of $A$. Then the **image** of $C$ under $f$, denoted by $f(C)$, is the set $f(C) = \{b \in B : b = f(c) \text{ for some } c \in C\}$. Thus, $f(C)$ consists of all those elements which can be obtained by applying $f$ to some element of $C$.

## Example 4

If $f : D_4 \to S_4$ is the homomorphism displayed above, then the image $f(D_4)$ is

$$\{\text{id}, (1\,2\,3\,4), (1\,3)(2\,4), (1\,4\,3\,2), (1\,4)(2\,3), (2\,4), (1\,2)(3\,4), (1\,3)\}.$$

## Example 5

Let $n \geqslant 2$, let $G$ be the symmetric group $S_n$ and consider the homomorphism $f : G \to \mathbb{R}\backslash\{0\}$ given by $f(\alpha) = \text{sgn}(\alpha)$ for all $\alpha \in S_n$. As sgn can take only the values $1, -1$, these form the image of $G$ under $f$. Thus, $f(G) = \{1, -1\}$, which is the cyclic subgroup of $\mathbb{R}\backslash\{0\}$ generated by $-1$.

## Example 6

Let $n > 2$ be an integer and let $f : S_n \to S_{n+1}$ be the function such that

$$f : \alpha = \begin{pmatrix} 1 & 2 & \dots & n \\ \alpha(1) & \alpha(2) & \dots & \alpha(n) \end{pmatrix} \mapsto \begin{pmatrix} 1 & 2 & \dots & n & n+1 \\ \alpha(1) & \alpha(2) & \dots & \alpha(n) & n+1 \end{pmatrix}.$$

Thus, $f(\alpha)$ has the same effect as $\alpha$ on $1, 2, \dots, n$ and fixes $n+1$. Then $f$ is a homomorphism and the image $f(S_n)$ is the set $\{\beta \in S_{n+1} : \beta(n+1) = n+1\}$, in other words the stabilizer of $n+1$.

## Example 7

Let $f : D_4 \to S_4$ be as above. Let $K$ be the cyclic subgroup $< r_1 >= \{e, r_1, r_2, r_3\}$ of $D_4$. Then

$$f(K) = \{f(e), f(r_1), f(r_2), f(r_3)\} = \{\text{id}, (1\ 2\ 3\ 4), (1\ 3)(2\ 4), (1\ 4\ 3\ 2)\}.$$

This is the cyclic subgroup generated by (1 2 3 4).

## Theorem 4

If $f : G \to H$ is a homomorphism and $K$ is a subgroup of $G$ then the image $f(K)$ is a subgroup of $H$. In particular, $f(G)$ is a subgroup of $H$.

PROOF

We use the subgroup criterion.

**SG1**: By Theorem 2(i), $e_H = f(e_G) \in f(K)$ so $e_H \in f(K)$ which is therefore non-empty.

**SG2**: Let $h_1, h_2 \in f(K)$. By the definition of image, $h_1 = f(k_1)$ and $h_2 = f(k_2)$ for some $k_1, k_2 \in K$. Thus, $h_1 h_2 = f(k_1) f(k_2) = f(k_1 k_2) \in f(K)$.

**SG3**: Let $h \in f(K)$. Then $h = f(k)$ for some $k \in K$. By Theorem 2(ii), $h^{-1} = (f(k))^{-1} = f(k^{-1}) \in f(K)$.
By the subgroup criterion, $f(K)$ is a subgroup of $H$.

## Example 8

Number the sides and vertices of the rectangle as shown.

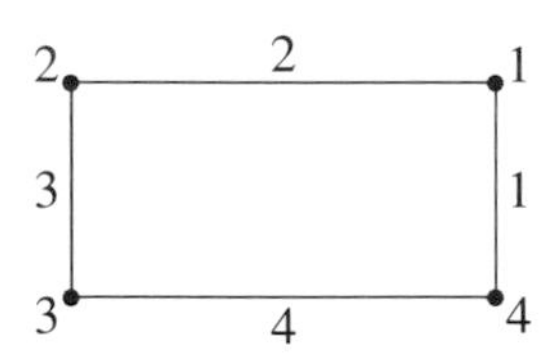


Let $G$ be the group of symmetries of the rectangle. Then $G$ acts on the vertices and the sides $\{1, 2, 3, 4\}$ so there are two homomorphisms $f_1 : G \to S_4$, arising from the vertices, and $f_2 : G \to S_4$, arising from the sides. Then

$$f_1(G) = \{\text{id}, (1\ 3)(2\ 4), (1\ 4)(2\ 3), (1\ 2)(3\ 4)\};$$

$$f_2(G) = \{\text{id}, (1\ 3)(2\ 4), (1\ 3), (2\ 4)\}.$$

These are different but both are isomorphic to Klein's 4-group. The fact that they are different shows that to say that two groups are isomorphic is not the same as saying they are equal, even when both are subgroups of the same bigger group. In fact, in Chapter 15 we shall see a difference between them when they are viewed as subgroups of $S_4$ rather than as groups in their own right.

If $f : G \to H$ is an *injective* homomorphism then $G \cong f(G)$ with $g \mapsto f(g)$. For example, $D_4$ is isomorphic to the subgroup $f(D_4) =$

$$\{\text{id}, (1\,2\,3\,4), (1\,3)(2\,4), (1\,4\,3\,2), (1\,4)(2\,3), (2\,4), (1\,2)(3\,4), (1\,3)\}$$

of $S_4$.

## Groups of matrices and linear transformations

We indicated in Chapter 5 that the group of symmetries of the circle and the group of $2 \times 2$ orthogonal matrices are closely related. As we show below, they are isomorphic. A similar situation arises for any subgroup $G$ of the general linear group $GL_2(\mathbb{R})$. Such a group is always isomorphic to a group of linear transformations.

Let $G$ be a subgroup of $GL_2(\mathbb{R})$. Then $G$ acts on $\mathbb{R}^2$ by the rule $A * x = Ax$ and the corresponding permutation of $\mathbb{R}^2$ is the linear transformation $f_A : x \mapsto Ax$. Hence, if $X = \mathbb{R}^2$, we get a homomorphism $f : G \to S_X$ where, for $A \in G$, $f(A) = f_A$. We shall see below that $f$ is injective, so $G$ is isomorphic to $f(G)$, which is a group of linear transformations. In particular, $GL_2(\mathbb{R})$ is isomorphic to the group of all linear transformations $f_A : \mathbb{R}^2 \to \mathbb{R}^2$ with $A$ invertible, as was discussed on page 43.

To check that the above homomorphism $f$ is injective we have to show that if $f_A = f_B$ then $A = B$. If $f_A = f_B$ then $f_A\left(\begin{pmatrix} 1 \\ 0 \end{pmatrix}\right) = f_B\left(\begin{pmatrix} 1 \\ 0 \end{pmatrix}\right)$ and $f_A\left(\begin{pmatrix} 0 \\ 1 \end{pmatrix}\right) = f_B\left(\begin{pmatrix} 0 \\ 1 \end{pmatrix}\right)$ and it follows easily that $A = B$.

### Example 9

In Chapter 5, we saw that the set of all orthogonal $2 \times 2$ matrices is a subgroup $H$ of $GL_2(\mathbb{R})$. By Theorem 4 of Chapter 3, the linear transformations corresponding to orthogonal $2 \times 2$ matrices are precisely the rotations $\text{rot}_\theta$ and reflections $\text{ref}_\theta$. We can now see that $H$ and $O_2$ are isomorphic under the isomorphism $A \mapsto f_A$. The name orthogonal group $O_2$ is used for both these groups.

### Example 10

Let $G = \left\{\begin{pmatrix} \cos\theta & -\sin\theta \\ \sin\theta & \cos\theta \end{pmatrix} : 0 \leqslant \theta < 2\pi\right\}$, which is a subgroup of $GL_2(\mathbb{R})$ consisting of the rotation matrices $A_\theta$. We saw in Section 3.3 that each $f_{A_\theta} = \text{rot}_\theta$. This group $G$ is isomorphic to the special orthogonal group $SO_2 = \{\text{rot}_\theta : 0 \leqslant \theta < 2\pi\}$. The same name, special orthogonal group $SO_2$, is used for both groups.

## EXERCISE 6

Let $g : D_4 \to S_4$ be the homomorphism arising from the action of $D_4$ on the sides of the square, numbered as in Fig 7.1. Find $g(a)$ for each $a \in D_4$. Let $f : D_4 \to S_4$ be the homomorphism arising from the action of $D_4$ on the vertices. Is $g = f$? Is $g(D_4) = f(D_4)$?

**EXERCISE 7**

Consider the action of $D_4$ on the diagonals of the square, labelled 1 and 2. For each of the eight elements $g$ of $D_4$ find the corresponding permutation $f_g$ of $\{1, 2\}$. Is the homomorphism given by $f : D_4 \to S_2$, where $g \mapsto f_g$, an injection?

## 9.4 Cayley's Theorem

We have seen that several groups, for example $D_4$, are isomorphic to subgroups of symmetric groups. It is reasonable to ask which groups are isomorphic to subgroups of symmetric groups $S_X$ for some set $X$. The answer, perhaps surprisingly, is *all* groups. This was first observed by **A. Cayley** and we state it as a theorem bearing his name. We shall prove Cayley's Theorem using an action of $G$ on itself. For $g, x \in G$, let $g * x = gx \in G$. We check that this gives an action. **GA1** holds because $e * x = ex = x$. For **GA2**, let $g, h, x \in G$. By associativity, $g * (h * x) = g(hx) = (gh)x = (gh) * x$. So we do have an action of $G$ on itself. Before the theorem, we give an example to illustrate the method of proof.

### Example 11

Let $G = \{e, r, s, t\}$ be the group of symmetries of the rectangle as given in Fig 4.1. Number the elements $e = g_1,\ r = g_2,\ s = g_3$ and $t = g_4$. Then $r * g_1 = rg_1 = g_2,\ r * g_2 = g_1,\ r * g_3 = g_4$ and $r * g_4 = g_3$. Thus, the permutation $f_r$ of $G$ arising from the action of $r$ is $(1\ 2)(3\ 4)$. Similarly $f_e = \text{id}$, $f_s = (1\ 3)(2\ 4)$ and $f_t = (1\ 4)(2\ 3)$. The homomorphism $f : G \to S_4$, where $f : g \mapsto f_g$, is injective and $G \cong H$ where $H$ is the image $f(G) = \{\text{id}, (1\ 2)(3\ 4), (1\ 3)(2\ 4), (1\ 4)(2\ 3)\}$. Both $G$ and $H$ are familiar copies of Klein's 4-group.

### Theorem 5 (Cayley's Theorem)

Every group $G$ is isomorphic to a subgroup of $S_G$. In particular, if $G$ is finite then $G$ is isomorphic to a subgroup of $S_n$, where $n = |G|$.

PROOF

Arising from the action of $G$ on itself by the rule $g * x = gx$, there is a homomorphism $f : G \to S_G$ where, for $g \in G$, $f(g) = f_g$ is the permutation of $G$ with $f_g(x) = g * x = gx$. This homomorphism is injective for if $f_g = f_h$ then $g = ge = f_g(e) = f_h(e) = he = h$. Hence, $G$ is isomorphic to its image $f(G)$, which is a subgroup of $S_G$. If $G$ is finite of order $n$ then, as in the above example, we identify $S_G$ with $S_n$ by numbering the elements of $G$ from 1 to $n$.

**EXERCISE 8**

Number the eight elements of $D_4$ from 1 to 8 in the order used in the Cayley table, Table 1.2. Find the image in $S_8$ of the homomorphism $f : D_4 \to S_8$ arising from the action of $D_4$ on itself by $g * x = gx$. (For example, from the row of the table for left multiplication by $r_1$, $f(r_1) = (1\ 2\ 3\ 4)(5\ 6\ 7\ 8)$.)

## 9.5 Cyclic Groups

Let $n$ be a positive integer. Here we show that any two cyclic groups of order $n$, for example the rotation group of the regular $n$-gon and the group of $n$th roots of unity, are isomorphic. We do this by showing that any cyclic group of order $n$ is isomorphic to the group $Z_n$ under addition modulo $n$. It follows that any two cyclic groups of order $n$ are isomorphic to each other.

### ● Theorem 6

Let $G$ be a finite cyclic group and let $n = |G|$. Then $G$ is isomorphic to the group $\mathbb{Z}_n$ under addition modulo $n$. ●

PROOF

Since $G$ is cyclic of order $n$, it has a generator, $g$, say, which has order $n$. Then every element of $G$ can be written as $g^m$ for some $m \in \mathbb{Z}$. We saw on page 89 that, for $i, j \in \mathbb{Z}$, $g^i = g^j$ if and only if $\overline{i} = \overline{j}$. Hence, there is a bijective function $f : G \to \mathbb{Z}_n$ defined by

$$f : g^m \mapsto \overline{m}, \text{ with } f^{-1} : \overline{m} \mapsto g^m.$$

To show that $f$ is a homomorphism and hence an isomorphism, let $h, k \in G$. Remember that $Z_n$ is in additive notation so we have to check that $f(hk) = f(h) + f(k)$. Now $h = g^a$, $k = g^b$ for some integers $a, b$ and

$$f(hk) = f(g^a g^b) = f(g^{a+b}) = \overline{a+b} = \overline{a} + \overline{b} = f(h) + f(k).$$

Thus, $f$ is an isomorphism and $G \cong \mathbb{Z}_n$. ●

This theorem tells us that all finite cyclic groups of order $n$ are isomorphic to each other. This allows us to use the notation $C_n$ to stand for a cyclic group generated by an element of order $n$.

We now turn to infinite cyclic groups. These are all isomorphic to each other.

### ● Theorem 7

Let $G$ be an infinite cyclic group. Then $G$ is isomorphic to $(\mathbb{Z}, +)$, the integers under addition. ●

PROOF

Since $G$ is cyclic, it has a generator $g$, say. Since $G$ is infinite, $g$ has infinite order. Let $f : \mathbb{Z} \to G$ be the function defined by $f : n \mapsto g^n$. Since $G$ is cyclic, $f$ is surjective. Suppose that $g^p = g^q$. Then $g^{p-q} = e_G$. Since $g$ has infinite order this is only possible if $p - q = 0$, that is $p = q$. Thus, $f$ is injective as well as surjective and so it is bijective. Now, for $p, q \in \mathbb{Z}$

$$f(p+q) = g^{p+q} = g^p g^q = f(p)f(q)$$

and so $f$ is an isomorphism. ●

This tells us that any two infinite cyclic groups are isomorphic. We use the notation $C_\infty$ to stand for a cyclic group generated by an element of infinite order.

## EXERCISE 9

Let $G$ be a cyclic group and let $H$ be a group isomorphic to $G$. Show that $H$ is cyclic.

## EXERCISE 10

(This exercise tells us that all groups of order two are isomorphic to each other as are all groups of order three.)

(i) Let $G = \{e, g\}$ be a group of order two. Show that $g$ has order two and that $G$ is cyclic generated by $g$.

(ii) Let $G = \{e, g, h\}$ be a group of order three. Use the cancellation laws (or Latin square property of the Cayley table) to show that $gh \neq g$ and $gh \neq h$. Hence, show that $gh = e$, that $g^2 = h$ and that $G$ is cyclic generated by $g$.

(This does not extend to groups of order four as we have seen Klein's 4-group which is not cyclic. As an exercise you might like to try to show that any group of order four is either cyclic or isomorphic to Klein's 4-group. However, this will be easier once we have proved Lagrange's Theorem in Chapter 10. This famous theorem also gives an easy proof that for any *prime* number $p$ all groups of order $p$ are cyclic. Given Lagrange's Theorem, the proof of this is no harder than the proof for the special case $p = 3$ in the exercise.)

# 9.6 Further Exercises on Chapter 9

## EXERCISE 11

Let $G = (\mathbb{R}, +)$ and let $H$ be the orthogonal group $O_2$. Show that the function $f : G \to H$ given by $\theta \mapsto \text{rot}_\theta$ is a homomorphism. Is it an isomorphism?

## EXERCISE 12

Let $G = (\mathbb{R}, +)$ and let $H$ be the orthogonal group $O_2$. Is the function $f : G \to H$ given by $\theta \mapsto \text{ref}_\theta$ a homomorphism?

## EXERCISE 13

Using Table 1.1, show that the function $f : O_2 \to O_2$ given by the rules

$$f : \text{rot}_\theta \mapsto \text{rot}_{2\theta}; \quad f : \text{ref}_\theta \mapsto \text{ref}_{2\theta}$$

is a homomorphism. Is $f$ injective? Is $f$ surjective?

## EXERCISE 14

Check that for all rotations $\mathrm{rot}_\theta \in D_6$ and all reflections $\mathrm{ref}_\theta \in D_6$, the rotation $\mathrm{rot}_{2\theta} \in D_3$ and the reflection $\mathrm{ref}_{2\theta} \in D_3$. Show that the function $f : D_6 \to D_3$ given by the rules

$$f : \mathrm{rot}_\theta \mapsto \mathrm{rot}_{2\theta}; \qquad f : \mathrm{ref}_\theta \mapsto \mathrm{ref}_{2\theta}$$

is a homomorphism. Is $f$ injective? Is $f$ surjective?

## EXERCISE 15

Let $n$ be a positive integer. Show that the function $f : \mathbb{Z} \to \mathbb{Z}_n$ given by $a \mapsto \overline{a}$ is a homomorphism. (Remember that these groups are in additive notation.) Is $f$ injective? Is $f$ surjective?

## EXERCISE 16

Replace the word 'isomorphism' by 'homomorphism' in Theorem 3. Does the theorem remain true? If not, what can you say about the orders of $g$ and $f(g)$? What happens if $f$ is injective?

## EXERCISE 17

Let $f : G \to H$ be a homomorphism and let $g \in G$. Show that $f(< g >) =< f(g) >$, the cyclic subgroup of $H$ generated by $f(g)$.

## EXERCISE 18

(i) Find three different subgroups of $S_4$ isomorphic to $D_4$ by numbering the vertices of the square in different ways.

(ii) Use the action of $D_4$ on the nine numbered squares to find a subgroup of the symmetric group $S_9$ isomorphic to $D_4$ and containing $(1\ 7\ 9\ 3)(2\ 4\ 8\ 6)$.

## EXERCISE 19

Find a subgroup of $S_6$ isomorphic to $D_3$ and containing $(1\ 2\ 3)(4\ 5\ 6)$. (Hint: Cayley's Theorem.)

# 10 • Cosets and Lagrange's Theorem

The main theorem of this chapter, **Lagrange's** Theorem, states that if $H$ is a subgroup of a finite group $G$ then the order of $H$ is a factor of the order of $G$. We shall prove this by finding a partition of $G$ into subsets, called left cosets, each of which has the same number of elements as $H$. As a consequence, we shall see that every group of prime order is cyclic.

## 10.1 Left Cosets

Let $G$ be a group and $H$ be a subgroup of $G$. Let $g \in G$. The set $gH = \{gh : h \in H\}$ of products of $g$ with elements of $H$, with $g$ on the left, is called a **left coset** of $H$ in $G$.

### Example 1

Let $G$=$D_4$ and let $H$ be the cyclic subgroup $< s_4 >= \{e, s_4\}$. Then

$$\begin{array}{lllll} eH & = & \{ee, es_4\} & = & \{e, s_4\}, \\ r_1H & = & \{r_1e, r_1s_4\} & = & \{r_1, s_1\}, \\ r_2H & = & \{r_2e, r_2s_4\} & = & \{r_2, s_2\}, \\ r_3H & = & \{r_3e, r_3s_4\} & = & \{r_3, s_3\}, \\ s_1H & = & \{s_1e, s_1s_4\} & = & \{s_1, r_1\}, \\ s_2H & = & \{s_2e, s_2s_4\} & = & \{s_2, r_2\}, \\ s_3H & = & \{s_3e, s_3s_4\} & = & \{s_3, r_3\}, \\ s_4H & = & \{s_4e, s_4s_4\} & = & \{s_4, e\}. \end{array}$$

Notice various aspects of this example.

- $g$ always occurs in $gH$. This is simply because $e \in H$ so $g = ge \in gH$.
- It is possible to have $g_1H = g_2H$ with $g_1 \neq g_2$. Indeed, each left coset appears twice in the above list. For example $\{r_1, s_1\}$ appears as both $r_1H$ and $s_1H$. The four different left cosets which occur are $r_1H$, $r_2H$, $r_3H$, and $eH$.
- The different left cosets form a partition of the group $G$.
- $H$ itself appears as the coset $eH$.
- Each left coset contains two elements, the same number as $H$.
- The order of $G$ = the number of different left cosets of $H$ $\times$ the order of $H$.
- Each of the cosets of $H$ is equal to one of the subsets $\text{send}_1(y)$, for some $y$ in the orbit of 1 for the action of $D_4$ on the nine squares.

| 1 | 2 | 3 |
|---|---|---|
| 4 | 5 | 6 |
| 7 | 8 | 9 |

$$\begin{aligned} \text{send}_1(1) &= \{e, s_4\} &= eH \\ \text{send}_1(7) &= \{r_1, s_1\} &= r_1H \\ \text{send}_1(9) &= \{r_2, s_2\} &= r_2H \\ \text{send}_1(3) &= \{r_3, s_3\} &= r_3H \end{aligned}$$

Each of these occurs more generally. We begin with the last one.

## Theorem 1

Let $G$ be a group acting on a set $X$, let $x \in X$ and let $H = \text{stab}(x)$. Let $g \in G$ and let $y = g * x$. Then $gH = \text{send}_x(y)$.

PROOF

Let $k \in \text{send}_x(y)$. Thus, $k * x = y$. We wish to show that $k \in gH$. Now

$$g^{-1}k * x = g^{-1} * (k * x) = g^{-1} * y = g^{-1} * (g * x) = (g^{-1}g) * x = e * x = x.$$

Hence, $g^{-1}k \in \text{stab}(x) = H$ and so $k = g(g^{-1}k) \in gH$. Thus, $\text{send}_x(y) \subseteq gH$.

Now let $k \in gH$. Then $k = gh$ for some $h \in H = \text{stab}(x)$. Thus

$$k * x = gh * x = g * (h * x) = g * x = y$$

and so $k \in \text{send}_x(y)$. This shows that $gH \subseteq \text{send}_x(y)$ and, as we already have the reverse inclusion, we have $gH = \text{send}_x(y)$.

## EXERCISE 1

Let $G = D_3$ and let $H = \{e, s_1\}$. Calculate the left coset $gH$ for each $g \in G$.

## EXERCISE 2

In the action of the orthogonal group on the set of lines through the origin, the stabilizer of the $x$-axis is $H = \{\text{id}, \text{rot}_\pi, \text{ref}_0, \text{ref}_\pi\}$, a copy of Klein's 4-group. Using Table 1.1, compute all the elements of the left coset $gH$ where $g = \text{rot}_{\frac{\pi}{4}}$. Hence, write down $\text{send}_L(L')$ where $L$ is the $x$-axis and $L'$ is the line $y = x$.

## EXERCISE 3

Consider the action of the dihedral group $D_4$ on the four triangles shown.

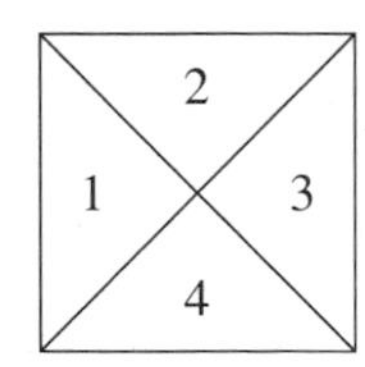


Find stab(1) and $\text{send}_1(2)$. Express $\text{send}_1(2)$ as a left coset of stab(1).

## 10.2 Left Cosets as Equivalence Classes

In Section 10.1, we have seen an example where the left cosets of a subgroup $H$ form a partition of a group $G$. This suggests that there is an underlying equivalence relation on $G$ whose equivalence classes are the left cosets of $H$. Once we have shown that this is always true, the theory of equivalence relations will give basic information on left cosets, for example, when two cosets $g_1 H$ and $g_2 H$ are equal.

Let $G$ be a group, let $H$ be a subgroup of $G$ and let $a, b \in G$. We say that $a$ **is equivalent to** $b$ **modulo** $H$, and write $a \equiv b \bmod H$, if $b^{-1}a \in H$. Below, we justify the use of the word 'equivalence' here by showing that this is indeed an equivalence relation.

### Example 2

For a group in additive notation, the condition $b^{-1}a \in H$ becomes $a - b \in H$. For example, let $G = \mathbb{Z}$ and $H = 5\mathbb{Z}$, the subgroup consisting of all the multiples of 5. Then for $a, b \in \mathbb{Z}$, $a \equiv b \bmod H$ precisely when $a - b \in H$, that is when $a - b$ is a multiple of 5. Thus, when $H = 5\mathbb{Z}$, equivalence modulo $H$ is exactly the same as equivalence mod 5 as described in Chapter 8.

### Example 3

(i) If $G = D_4$ and $H = \langle s_4 \rangle$ then $r_3^{-1}s_3 = s_4 \in H$ so $s_3 \equiv r_3 \bmod H$.

(ii) Let $G = O_2$, the orthogonal group, and let $H$ be the subgroup of rotations. Then $\text{ref}_\alpha \equiv \text{ref}_\beta \bmod H$ for any $\alpha$ and $\beta$ since, using Table 1.1, $\text{ref}_\beta^{-1}\,\text{ref}_\alpha = \text{ref}_\beta\,\text{ref}_\alpha = \text{rot}_{\beta-\alpha}$. Similarly, any two rotations are equivalent modulo $H$. However a rotation $\text{rot}_\alpha$ and a reflection $\text{ref}_\beta$ cannot be equivalent modulo $H$ because $(\text{rot}_\alpha)^{-1}\,\text{ref}_\beta$ is a reflection.

### Theorem 2

Let $G$ be a group and let $H$ be a subgroup of $G$. Then equivalence modulo $H$ is an equivalence relation on $G$.

PROOF

We must prove that the relation is reflexive, symmetric and transitive.

- Reflexive: if $a \in G$ then $a^{-1}a = e \in H$ and so $a \equiv a \bmod H$.
- Symmetric: let $a, b \in G$ be such that $a \equiv b \bmod H$. Then $b^{-1}a \in H$. Now $a^{-1}b = (b^{-1}a)^{-1} \in H$ and so $b \equiv a \bmod H$.
- Transitive: let $a, b, c \in G$ be such that $a \equiv b \bmod H$ and $b \equiv c \bmod H$. Then $b^{-1}a \in H$ and $c^{-1}b \in H$, so $c^{-1}a = c^{-1}bb^{-1}a \in H$. Thus, $a \equiv c \bmod H$.

Since we have shown that equivalence mod $H$ is always an equivalence relation, we know from Theorem 3 of Chapter 8 that the corresponding equivalence classes partition $G$. Our next step is to show that the equivalence classes are the left cosets. Remember that the equivalence class $\overline{a}$ is defined to be $\{b \in G : b \equiv a \bmod H\} = \{b \in G : a^{-1}b \in H\}$.

## • Theorem 3

Let $G$ be a group, $H$ be a subgroup of $G$, and $a \in G$. The equivalence class $\overline{a}$ of $a$ under equivalence modulo $H$ is the left coset $aH$. Hence the left cosets of $H$ in $G$ form a partition of $G$. •

PROOF

Firstly let $c \in aH$. Then $c = ah$ for some $h \in H$ and so $a^{-1}c = a^{-1}ah = h \in H$ and $c \equiv a \bmod H$. Thus, $c \in \overline{a}$ for all $c \in aH$ and therefore $aH \subseteq \overline{a}$. Now let $d \in \overline{a}$. Then $a^{-1}d \in H$ and so $a^{-1}d = h$ for some $h \in H$. Then $d = aa^{-1}d = ah \in aH$. Thus, $\overline{a} \subseteq aH$ and, as we already have the reverse inclusion, $\overline{a} = aH$ as required. It follows from Theorem 3 of Chapter 8 that the left cosets of $H$ in $G$ form a partition of $G$. •

## • Theorem 4

Let $G$ be a group, $H$ be a subgroup of $G$ and $a, b \in G$. Then

(i) $aH = bH$ if and only if $b^{-1}a \in H$;

(ii) $eH = H$;

(iii) $aH = H$ if and only if $a \in II$;

(iv) if $a \in bH$ then $aH = bH$. •

PROOF

(i) We know from Theorem 1 of Chapter 8 that $\overline{a} = \overline{b}$ if and only if $b^{-1}a \in H$. But by Theorem 3, $\overline{a} = aH$ and $\overline{b} = bH$. Hence, $aH = bH$ if and only if $b^{-1}a \in H$.

(ii) Because $eh = h$ for all $h \in H$, it is clear that $eH = H$.

(iii) By (ii), $aH = H$ if and only if $aH = eH$. It follows on setting $b = e$ in (i) that $aH = H$ if and only if $a \in H$.

(iv) Suppose that $a = bh \in aH$. Then $b^{-1}a = h \in H$ and, by (i), $aH = bH$. •

This theorem saves time in the calculation of all the left cosets of a given subgroup. For example, return to Section 10.1 where $G = D_4$ and $H = < s_4 >$. The first coset we listed was $H = eH$. As $s_4 \in H$, there is no need to calculate $s_4H$ which must also equal $H$. The next coset we found was $r_1H = \{r_1, s_1\}$. By Theorem 4(iv), there is now no need to calculate $s_1H$ which must equal $r_1H$. Similarly, once we have $r_2H = \{r_2, s_2\}$ and $r_3H = \{r_3, s_3\}$ we know that $s_2H = r_2H$ and $s_3H = r_3H$.

## EXERCISE 4

Let $G = D_4$ and let $H = \{e, r_1, r_2, r_3\}$. How many distinct left cosets of $H$ in $G$ are there? How many elements are in each?

## 10.3 Lagrange's Theorem

In the examples and exercises on left cosets you will have noticed that the left cosets are the same size as the relevant subgroup.

### ● *Theorem 5*

Let $G$ be a group and let $H$ be a subgroup of $G$. Let $a \in G$. There is a bijective function $\theta : H \rightarrow aH$ given by $\theta(h) = ah$ for all $h \in H$. Hence $|aH| = |H|$. ●

PROOF

Since $G$ is a group, $a$ has an inverse $a^{-1}$. The function $\phi : aH \rightarrow H$ given by $\phi(ah) = a^{-1}(ah)$ is clearly an inverse of $\theta$. By Theorem 1 of Chapter 2, $\theta$ is bijective. ●

We now have all the ingredients for the proof of Lagrange's Theorem.

### ● *Theorem 6 (Lagrange's Theorem)*

Let $G$ be a finite group and let $H$ be a subgroup of $G$. Then the order of $H$ is a factor of the order of $G$. More precisely, $|G| = m|H|$ where $m$ is the number of different left cosets of $H$ in $G$. ●

PROOF

Let $|G| = n$ and let $|H| = k$. By Theorem 3, the left cosets of $H$ in $G$ form a partition of $G$. By Theorem 5, the number of elements in each left coset is $k$. Since the left cosets partition $G$, each element of $G$ is in exactly one of the $m$ left cosets. Hence, $|G| = k + k + \ldots + k = mk = m|H|$. ●

### *Example 4*

The symmetric group $S_4$ cannot have a subgroup of order 5 since $|S_4| = 24$ which is not an exact multiple of 5.

Let $G$ be a finite group of order $n$. Let $a \in G$. What can we say about the order of $a$?

### ● *Theorem 7*

Let $G$ be a finite group and let $a \in G$. Then

(i) the order of $a$ is a factor of $|G|$;

(ii) $a^{|G|} = e$. ●

PROOF

Let the order of $a$ be $k$.

(i) The cyclic subgroup $< a >$ of $G$ has order $k$ by Theorem 3(ii) of Chapter 6. By Lagrange's Theorem, $k$ is a factor of $|G|$.

(ii) By (i), $k$ divides $n$, so $n = kq$ for some $q$. By Theorem 3(iii) of Chapter 6, $a^{|G|} = e$. ●

**EXERCISE 5**

(This exercise shows that any non-cyclic group of order 4 is isomorphic to Klein's 4-group.) Let $G = \{e, a, b, c\}$ be a group of order 4 which is not cyclic.

(i) What are the orders of the non-identity elements $a$, $b$ and $c$?

(ii) Use the cancellation laws or the Latin square property to show that $ab = c$ and complete the Cayley table for $G$.

## 10.4 Consequences of Lagrange's Theorem

The next theorem is a direct consequence of Lagrange's Theorem and demonstrates its strength. It shows that, for each prime number $p$, all groups of order $p$ are cyclic and hence isomorphic to each other.

### Theorem 8

Let $p$ be a prime number and let $G$ be a group of order $p$.

(i) The only subgroups of $G$ are $G$ and $\{e\}$.

(ii) $G$ is cyclic. ●

PROOF

(i) Let $H$ be a subgroup of $G$. By Lagrange's Theorem, the order of $H$ must be a factor of $p$. Since $p$ is prime the only possibilities are $|H| = p$, in which case $H = G$, and $|H| = 1$, in which case $H = \{e\}$.

(ii) Let $g \in G$ with $g \neq e$. Let $H = <g>$ be the cyclic subgroup generated by $g$. Since $g \neq e$ we know that $H \neq \{e\}$ and so, by (i), $H = G$ and $G = <g>$ is cyclic. ●

Methods based on Lagrange's Theorem can also be used to obtain information about groups of non-prime order.

### Example 5

Consider the symmetric group $S_3$ which has order 6. By Lagrange's Theorem, the possible orders for proper non-trivial subgroups are 2 and 3. As these are both prime, any such subgroup must be cyclic. We can obtain a list of them by working out the cyclic subgroups generated by each element in turn. Each of the three transpositions $t$ generates a subgroup $<t> = \{t, \text{id}\}$. The two 3-cycles, (1 2 3) and (1 3 2) generate the same cyclic subgroup $\{\text{id}, (1\ 2\ 3), (1\ 3\ 2)\}$. Thus, including the trivial subgroup $<\text{id}>$ and the whole group, $S_3$ has six subgroups.

## • Theorem 9

Let $H$ and $K$ be subgroups of a finite group $G$.

(i) If $|H|$ and $|K|$ are coprime then $H \cap K = \{e\}$.

(ii) If $H$ and $K$ are distinct and both have prime order $p$ then $H \cap K = \{e\}$. •

PROOF

(i) By Theorem 3 of Chapter 5, the intersection $H \cap K$ is a subgroup of $G$. As it is contained in $H$ and in $K$, it is also a subgroup of $H$ and a subgroup of $K$. By Lagrange's Theorem, its order is a common factor of $|H|$ and $|K|$, and so must be 1. Hence, $H \cap K = \{e\}$.

(ii) Here, $H \cap K$ is a subgroup of $H$ which has prime order so, by Theorem 8, either $H \cap K = H$ or $H \cap K = \{e\}$. If $H \cap K = H$ then $H \subseteq K$ but this is impossible because $|H| = |K|$ and $H \neq K$. Hence, $H \cap K = \{e\}$. •

## Example 6

Let $G$ be a group of order 15. The possible orders of elements or subgroups of $G$ are 1,3,5 and 15. We shall show that $G$ must have an element of order 3. To do this, we first show that the number of elements of order 5 is a multiple of 4. If $g \in G$ has order 5 then $g$ is in the cyclic subgroup $< g >= \{e, g, g^2, g^3, g^4\}$ which contains four elements of order 5 together with $e$. By Theorem 9(ii), no element of order 5 is in two different cyclic subgroups of order 5. Thus, each element of order 5 is in exactly one cyclic subgroup of order 5. Each such subgroup contains four elements of order 5. It follows that if the number of different subgroups of order 5 in $G$ is $d$ then the total number of elements of order 5 in $G$ is $4d$. As $|G| = 15$, it has at most 12 elements of order 5. Only $e$ has order 1 so there must be at least one element of order 15 or 3. But if $g$ has order 15, then $g^5$ has order 3 and so $G$ must have an element of order 3.

Notice that the method we used to show the existence of an element of order 3 cannot be applied to show that $G$ has an element of order 5. This is because, whereas 14 is not a multiple of $(5 - 1)$, it is a multiple of $(3 - 1)$ so it appears feasible that $G$ could have 14 elements of order 3 and seven subgroups of order 3. We shall see later that this too is impossible.

## • Theorem 10

Let $G$ be a group and let $a, b \in G$ with finite orders $n$ and $m$ respectively. If $ba = ab$ and $n$ and $m$ are coprime then $ab$ has order $mn$. •

PROOF

By Theorem 9(i) with $H =< a >$ and $K =< b >$, the intersection $< a > \cap < b >$ is $\{e\}$. By Theorem 5 of Chapter 6, the order of $ab$ is the l.c.m. of $m$ and $n$ which is $mn$. •

## EXERCISE 6

Let $G$ be a group of order 21. Show that all proper subgroups of $G$ are cyclic and that $G$ must have an element of order 3.

## 10.5 Applications to Number Theory

Earlier, we promised some applications to number theory. We obtain these by applying ideas of order to the group $U(\mathbb{Z}_p)$ for a prime number $p$. This allows us to compute the remainder when a number of the form $a^b$ is divided by $p$. For example, let us compute the remainder on dividing $100^{100}$ by 7. As $100 \equiv 2 \bmod 7$, $\overline{100} = \overline{2}$ in $U(\mathbb{Z}_7)$. It is easy to check that $\overline{2}$ has order 3 in $U(\mathbb{Z}_7)$ and so, by Theorem 3(i) of Chapter 6, $\overline{2}^{100} = \overline{2}$. Hence, $\overline{100}^{100} = \overline{2}^{100} = \overline{2}$. In other words, the remainder on dividing $100^{100}$ by 7 is 2.

We now apply Theorem 7 to obtain a general result on remainders. This result is known as **Fermat's** Little Theorem and now has an important role in cryptography and primality testing.

### • Theorem 11 (Fermat's Little Theorem)

Let $p$ be a prime number and let $a$ be an integer. Then $a^p \equiv a \bmod p$. •

PROOF

The group $U(\mathbb{Z}_p)$ has order $p - 1$ by Theorem 4 of Chapter 8

If $a$ is not a multiple of $p$, the same theorem tells us that $\overline{a} \in U(\mathbb{Z}_p)$. By Theorem 7, $(\overline{a})^{p-1} = e_G = \overline{1}$. In other words, $a^{p-1} \equiv 1 \bmod p$, that is $a^{p-1} - 1$ is a multiple of $p$. Therefore, $a^p - a = a(a^{p-1} - 1)$ is also a multiple of $p$, that is, $a^p \equiv a \bmod p$.

On the other hand, if $a$ is a multiple of $p$ then so is $a^p - a$. Thus, $a^p \equiv a \bmod p$ in this case also. •

The next theorem, Wilson's Theorem, is obtained by determining which of the $p - 1$ elements of $U(\mathbb{Z}_p)$ are their own inverses. For example, when $p = 11$, we have that $\overline{2}$ and $\overline{6}$ are inverses of each other, as are $\overline{3}$ and $\overline{4}$, $\overline{5}$ and $\overline{9}$, and $\overline{7}$ and $\overline{8}$. Only $\overline{1}$ and $\overline{10}$ are their own inverses. Multiplying all 10 elements together and pairing inverse pairs together, we get

$$\overline{1}(\overline{2}\,\overline{6})(\overline{3}\,\overline{4})(\overline{5}\,\overline{9})(\overline{7}\,\overline{8})\overline{10} = \overline{10}.$$

By associativity and commutativity in $\mathbb{Z}_p$, the left-hand side can be rewritten as $\overline{10!}$. Thus, $\overline{10!} = \overline{10} = \overline{-1}$, in other words, $10! \equiv -1 \bmod 11$.

### • Theorem 12 (Wilson's Theorem)

Let $p$ be a prime number. Then $(p - 1)! \equiv -1 \bmod p$. •

PROOF

Let $1 \leqslant a \leqslant p - 1$. If $a = 1$ or $p - 1$ then $a^2 \equiv 1 \bmod p$ so $\overline{a}$ is its own inverse in the group $U(\mathbb{Z}_p)$. Let $1 < a < p - 1$. Then $0 < a - 1 < p - 2$ and $2 < a + 1 < p$ so neither $a - 1$ nor $a + 1$ is a multiple of $p$. As $p$ is prime, the product $(a - 1)(a + 1) = a^2 - 1$ cannot be a multiple of $p$. Thus, $a^2 \not\equiv 1 \bmod p$ and so $\overline{a}$ is not its own inverse. Hence, only $\overline{1}$ and $\overline{p-1}$ are their own inverses. Pairing off the other elements in inverse pairs, as was done above for $p = 11$, we have that

$$\overline{(p-1)!} = \overline{p-1}\,\overline{1}\ldots\overline{1}\,\overline{1} = \overline{p-1} = \overline{-1}.$$

In other words, $(p - 1)! \equiv -1 \bmod p$. •

**EXERCISE 7**

Find the remainders when $297^{792}$ is divided by 5, 7 and 11.

## 10.6 Right Cosets

Throughout this chapter we have concentrated on the left cosets $gH = \{gh : h \in H\}$ of a subgroup $H$ of a group $G$. We can also define the **right coset** $Hg$ which is the set $\{hg : h \in H\}$. The right cosets behave similarly to the left cosets.

- $He = H$.
- The right cosets form a partition of $G$ with $g = eg \in Hg$.
- $Ha = Hb$ if and only if $ab^{-1} \in H$. Combining this with the left-handed version in Theorem 4(i), $Ha = Hb$ if and only if $a^{-1}H = b^{-1}H$.
- There is a bijection from $H$ to $Hg$ given by $h \mapsto hg$ so that $|Hg| = |H|$.
- If $G$ is finite then $|G| = j|H|$ where $j$ is the number of distinct right cosets of $H$ in $G$.
- When $G$ is finite the number of distinct left cosets and the number of distinct right cosets are equal, both being $\frac{|G|}{|H|}$. In general there are inverse functions between the set of left cosets and the set of right cosets given by

$$aH \mapsto Ha^{-1} \text{ and } Hb \mapsto b^{-1}H.$$

**EXERCISE 8**

Let $H$ be the cyclic subgroup $< s_4 >$ of the dihedral group $D_4$.

(i) Calculate the right cosets $Hr_1$ and $Hr_2$.

(ii) Show that $Hr_1 \neq r_1H$ but that $Hr_2 = r_2H$.

(iii) We know from Section 10.1 that $r_1H = s_1H$. Show that $Hr_1 \neq Hs_1$ but that $Hr_1^{-1} = Hs_1^{-1}$.

(iv) Show that the right coset $Hr_1$ of $H$ is equal to the left coset $r_1K$ of the subgroup $K =< s_2 >$.

## 10.7 Further Exercises on Chapter 10

**EXERCISE 9**

Let $G$ be the orthogonal group and let $H = SO_2$ be the subgroup of rotations.

(i) Show that if $a = \text{rot}_\theta$ is a rotation then every element of $aH$ is a rotation. Show also that every rotation $\text{rot}_\beta$ is in $aH$. (Thus, the left coset $aH$ is the set of all rotations.)

(ii) Show that if $a = \text{ref}_\theta$ is a reflection then every element of $aH$ is a reflection. Show also that every reflection $\text{ref}_\beta$ is in $aH$. (Thus, the left coset $aH$ is the set of all reflections.)

## EXERCISE 10

Let $G$ be a group of order 105 and let $H$ be a subgroup of $G$. Show that if $|H| \geqslant 36$ then $H = G$.

## EXERCISE 11

Let $G$ be a group with two subgroups $H$ and $K$. Suppose that $|H| = 75$ and $|K| = 57$. Show that $H \cap K$ is cyclic.

## EXERCISE 12

Let $G$ be a group of order 35. Show that $G$ has an element of order 5 and an element of order 7. Hence show that if $G$ is abelian then $G$ is cyclic.

## EXERCISE 13

Let $H$ and $K$ be the subgroups of the symmetric group $S_5$ generated by $(1\ 2\ 3\ 4\ 5)$ and $(1\ 2\ 3\ 4)$ respectively. Show that $H \cap K = \{\text{id}\}$.

## EXERCISE 14

Let $H$ be a subgroup of the symmetric group $S_4$. Show that if $|H| > 8$ then $|H| \geqslant 12$.

## EXERCISE 15

Let $p$ be a polynomial in four variables. Show that if 13 permutations stabilize $p$ then $p$ is symmetric.

## EXERCISE 16

The aim of this exercise is to show that the converse of Fermat's Little Theorem is not true. That is, there exist non-prime numbers $q$ such that $a^q \equiv a \bmod q$ for all integers $a$. Numbers that have this property and are not, in fact, prime are known as Carmichael numbers. It has only recently been shown that there are infinitely many Carmichael numbers.

(i) For each of the primes $p = 3, 11, 17$, show that $a^{560} \equiv 1 \bmod p$ when $a$ is an integer which is not a multiple of $p$. (Hint: 560 is a multiple of $p - 1$ for each of these primes $p$.)

(ii) Show that, when $p = 3, 11$ or $17$, $p$ divides $a^{561} - a$ for each integer $a$.

(iii) Show that $a^{561} \equiv a \bmod 561$ for each integer $a$.

## EXERCISE 17

(This exercise shows that the converse of Wilson's Theorem is true.) Show that if $n$ is not a prime number and $n \neq 4$ then $(n-1)! \equiv 0 \bmod n$. Check that $(4-1)! \equiv 2 \bmod 4$.

## EXERCISE 18

Investigate the subgroups of the symmetric group $S_4$. For each factor $d$ of 24, can you find a subgroup of order $d$? How many subgroups of order $d$ can you find? Investigate the subgroups of the alternating group $A_4$.

# 11 • The Orbit-Stabilizer Theorem

We have seen in Chapter 10 that there is a close connection between the left cosets of a stabilizer and the subsets $\text{send}_x(y)$ arising from a group action. Every set of the form $\text{send}_x(y)$ is a left coset of the stabilizer. This allows us to apply general results about left cosets, including Lagrange's Theorem, to the subsets $\text{send}_x(y)$. In the last section, we derived a formula for the number of orbits of a group action.

## 11.1 The Orbit-Stabilizer Theorem

The first result concerns the number of elements in $\text{send}_x(y)$.

● *Theorem 1*

Let $G$ be a finite group acting on a set $X$ and let $x, y \in X$ with $y \in \text{orb}(x)$. Then $|\text{send}_x(y)| = |\text{stab}(x)|$ (that is the number of elements that send $x$ to $y$ is equal to the number of elements that send $x$ to itself). Hence, $|\text{send}_x(y)|$ is a divisor of $|G|$. ●

PROOF

Let $H = \text{stab}(x)$. By Theorem 1 of Chapter 10, $\text{send}_x(y)$ is a left coset of $H$ so, by Theorem 5 of Chapter 10, $|\text{send}_x(y)| = |H|$. By Lagrange's Theorem, this is a divisor of $|G|$. ●

In Section 10.1, we found the left cosets of the stabilizer $H$ of 1 for the action of $D_4$ on the nine numbered squares and observed that each of these was equal to $\text{send}_1(y)$ for some $y$ in the orbit of 1. This is shown in Table 11.1, where each element $y$ of the orbit of 1 is paired with a left coset. This pairing between the orbit of 1 and the set of left cosets

| | | |
|---|---|---|
| 1 | $\mapsto$ | $\text{send}_1(1) = \{e, s_4\} = eH = s_1H$ |
| 3 | $\mapsto$ | $\text{send}_1(3) = \{r_3, s_3\} = r_3H = s_3H$ |
| 5 | $\mapsto$ | $\text{send}_1(7) = \{r_1, s_1\} = r_1H = s_1H$ |
| 7 | $\mapsto$ | $\text{send}_1(9) = \{r_2, s_2\} = r_2H = s_2H$ |

**Table 11.1** Relationship between orb(1) and the left cosets of stab(1).

of $H$ in $G$, where $y$ is paired with $\text{send}_1(y)$, is performed by a bijective function $\text{send}_1$. Such a bijective function exists for the orbit of any element $x$ of a set $X$ when there is a group $G$ acting on $X$.

● *Theorem 2 (The Orbit-Stabilizer Theorem)*

Let $G$ be a group acting on a set $X$, let $x \in X$ and let $H = \text{stab}(x)$. Let $C(x)$ be the set of left cosets of $H$ in $G$. Then the function $\text{send}_x : \text{orb}(x) \to C(x)$ is a bijection. ●

PROOF

We need to show that the function $\text{send}_x$ is both injective and surjective. For surjectivity, let $gH \in C(x)$. Then $y = g * x \in \text{orb}(x)$ and $\text{send}_x(y) = gH$ by Theorem 1 of Chapter 10. Thus, $\text{send}_x$ is surjective. To see that it is also injective, let $y, z \in \text{orb}(x)$ be such that $\text{send}_x(y) = \text{send}_x(z)$. Let $g \in \text{send}_x(y) = \text{send}_x(z)$. Then $g * x = y$ because $g \in \text{send}_x(y)$ and $g * x = z$ because $g \in \text{send}_x(z)$. Hence, $y = g * x = z$ and therefore $\text{send}_x$ is injective. Since $\text{send}_x$ is both injective and surjective it is bijective. ●

## Example 1

We already have one bijection illustrating this theorem displayed in Table 11.1. For another example, consider the usual action of $O_2$ on $\mathbb{R}^2$. The orbit of the point $P = (1, 0)$ under this action is the unit circle. The stabilizer of $P$ is $\{\text{rot}_0, \text{ref}_0\}$. How does the bijection described in Theorem 2 work in this example? A typical member of $\text{orb}(P)$ is $(\cos\theta, \sin\theta)$ and $\text{send}_P((\cos\theta, \sin\theta)) = \{\text{rot}_\theta, \text{ref}_\theta\}$. The bijection in this case is therefore

$$(\cos\theta, \sin\theta) \longleftrightarrow \{\text{rot}_\theta, \text{ref}_\theta\} = \text{rot}_\theta \,\text{stab}(P).$$

## Example 2

Consider the symmetric group $S_3$ acting on the set $\{1, 2, 3\}$ by $\alpha * x = \alpha(x)$. Then $\text{stab}(2) = \{\text{id}, (1\ 3)\} = H$, say, and $\text{orb}(2) = \{1, 2, 3\}$. The bijection in this case, for $x = 2$, is shown below.

$$\begin{aligned} 2 &\longleftrightarrow \{\text{id}, (1\ 3)\} = H, \\ 1 &\longleftrightarrow \{(1\ 2), (1\ 3\ 2)\} = (1\ 2)H, \\ 3 &\longleftrightarrow \{(2\ 3), (1\ 2\ 3)\} = (2\ 3)H. \end{aligned}$$

When $G$ is a finite group, Lagrange's Theorem can be combined with the Orbit-Stabilizer Theorem to obtain a relationship between the sizes of an orbit and the corresponding stabilizer.

## Theorem 3 (Orbit-Stabilizer Theorem, finite form)

Let $G$ be a finite group acting on a set $X$ and let $x \in X$. Then

$$|\text{orb}(x)| \times |\text{stab}(x)| = |G|.$$ ●

PROOF

By the Orbit-Stabilizer Theorem, the number of elements in $\text{orb}(x)$ is the same as the number $m$ of left cosets of $\text{stab}(x)$. Applying Lagrange's Theorem with $H = \text{stab}(x)$, we obtain the required formula $|G| = m|H|$. ●

This result is the one that we hoped you would predict in your answers to exercises in Chapters 1,2 and 7. It shows that the size of an orbit $\text{orb}(x)$ is inversely proportional to the size of the stabilizer $\text{stab}(x)$.

## EXERCISE 1

The dihedral group $D_3$ acts on the four numbered equilateral triangles shown.

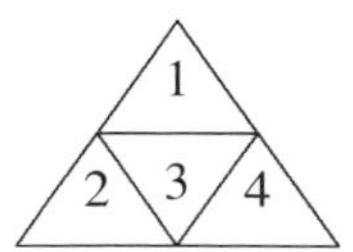


Display the bijection $\text{send}_2$ from the orbit of 2 to the set of left cosets of stab(2) as in Example 2. (With the triangle positioned as shown, rather than with the $x$-axis as a line of reflection, the three reflections in $D_3$ are $s_1 = \text{ref}_{\frac{\pi}{3}}$, $s_2 = \text{ref}_{\pi}$ and $s_3 = \text{ref}_{\frac{5\pi}{3}}$.)

# 11.2 Fixed Subsets

In our discussion of group actions on sets, we have concentrated on the size of orbits and stabilizers. In the next section, we give a formula for the number of orbits when both the group and set are finite. The formula involves the idea of fixed sets.

Let $G$ be a group acting on a set $X$. For each $g \in G$, the **fixed set** of $g$ is the subset $\text{fix}(g) = \{x \in X : g * x = x\}$ consisting of those elements of $X$ which are fixed by $g$.

For example, when the symmetric group $S_4$ acts on $X = \{1, 2, 3, 4\}$ the 4-cycle $f = (1\ 2\ 3\ 4)$ does not fix any element of $X$ so $\text{fix}(f) = \emptyset$. However, the transposition $t = (1\ 2)$ fixes 3 and 4 so $\text{fix}(t) = \{3, 4\}$.

Let $G$ be a finite group acting on a non-empty finite set $X$. Suppose we wish to count how many times we have $g * x = x$ for some $g \in G$ and some $x \in X$, in other words how many times we have an element of $G$ stabilizing or fixing an element of $X$. One way is to take each $x \in X$ in turn, compute the order of its stabilizer and then take the sum $\sum_{x \in X} |\text{stab}(x)|$ over all $x \in X$. Another way is to take each $g \in G$ in turn, ask how many elements of $X$ are fixed by $g$ and then take the sum $\sum_{g \in G} |\text{fix}(g)|$ over all $g \in G$.

Both approaches to this counting problem must give the same result. Thus

$$\sum_{g \in G} |\text{fix}(g)| = \sum_{x \in X} |\text{stab}(x)|,$$

both being the total number of pairs $g, x$ with $g \in G$, $x \in X$ and $g * x = x$.

We illustrate this by looking at the action of the dihedral group $D_4$ acting on the nine squares.

<table>
<tr><td>1</td><td>2</td><td>3</td></tr>
<tr><td>4</td><td>5</td><td>6</td></tr>
<tr><td>7</td><td>8</td><td>9</td></tr>
</table>

We first compute $\sum_{g \in G} |\operatorname{fix}(g)|$. The result is shown in Table 11.2 below. For example, the reflection $s_3$ in the $y$-axis fixes 2, 5 and 8 so $\operatorname{fix}(s_3) = \{2, 5, 8\}$. However, the rotation $r_1$ only fixes 5 so $\operatorname{fix}(r_1) = \{5\}$. The table shows that

$$\sum_{g \in G} \operatorname{fix}(g) = 9 + 1 + 1 + 1 + 3 + 3 + 3 + 3 = 24.$$

| $g$ | $\operatorname{fix}(g)$ | $\lvert\operatorname{fix}(g)\rvert$ |
|---|---|---|
| $e$ | $\{1, 2, 3, 4, 5, 6, 7, 8, 9\}$ | 9 |
| $r_1$ | $\{5\}$ | 1 |
| $r_2$ | $\{5\}$ | 1 |
| $r_3$ | $\{5\}$ | 1 |
| $s_1$ | $\{4, 5, 6\}$ | 3 |
| $s_2$ | $\{2, 5, 8\}$ | 3 |
| $s_3$ | $\{3, 5, 7\}$ | 3 |
| $s_4$ | $\{1, 5, 9\}$ | 3 |

| | |
|---|---|
| $e$ | rotation through 0. |
| $r_1$ | rotation through $\pi/2$. |
| $r_2$ | rotation through $\pi$. |
| $r_3$ | rotation through $3\pi/2$. |
| $s_1$ | reflection in the $x$-axis. |
| $s_2$ | reflection in the $y = x$ diagonal. |
| $s_3$ | reflection in the $y$-axis. |
| $s_4$ | reflection in the $y = -x$ diagonal. |

**Table 11.2** Sizes of fixed subsets when $D_4$ acts on nine squares.

We now take each $x \in X = \{1, 2, 3, 4, 5, 6, 7, 8, 9\}$ in turn and compute $\operatorname{stab}(x)$. Rather than take these in the order $1, 2, 3, \ldots$ we shall consider each orbit in turn. There are three orbits $O1 = \{1, 3, 9, 7\}$, $O2 = \{2, 6, 8, 4\}$ and $O3 = \{5\}$. Each square in $O1$ is stabilized only by id and one reflection, for example, $\operatorname{stab}(1) = \{\mathrm{id}, s_4\}$. Thus

$$\sum_{x \in O1} |\operatorname{stab}(x)| = 2 + 2 + 2 + 2 = 8.$$

The same is true for $O2$,

$$\sum_{x \in O2} |\operatorname{stab}(x)| = 2 + 2 + 2 + 2 = 8.$$

For the only element, 5, of $O3$, the stabilizer is the whole group so

$$\sum_{x \in O3} |\operatorname{stab}(x)| = 8.$$

Adding these subtotals for orbits together, we obtain the total for the whole of $X$,

$$\begin{aligned} \sum_{x \in X} |\operatorname{stab}(x)| &= \sum_{x \in O1} |\operatorname{stab}(x)| + \sum_{x \in O2} |\operatorname{stab}(x)| + \sum_{x \in O3} |\operatorname{stab}(x)| \\ &= 8 + 8 + 8 \\ &= 24 \\ &= \sum_{g \in G} |\operatorname{fix}(g)|. \end{aligned}$$

Notice that, for each orbit, the subtotal of the orders of the stabilizers is 8, the order of the group $D_4$. The overall total is then $3 \times 8$, that is (number of orbits) $\times$ (order of group). Also, the stabilizer is the same size for each element of a given orbit. These observations illustrate the theorems and proofs in the next section. You should notice similar behaviour in the next exercise.

## EXERCISE 2

Let $G$ be the dihedral group $D_3$ acting on the four triangles in Exercise 1. For each $g \in G$, find $\text{fix}(g)$ and hence obtain a list of all pairs $g, x$ with $g * x = x$. For each triangle $x$, find $\text{stab}(x)$ and hence obtain a second list of all pairs $g, x$ with $g * x = x$. Compare the two lists and check that $\sum_{g\in G} |\text{fix}(x)| = \sum_{x\in X} |\text{stab}(x)| =$ (number of orbits) $\times$ (order of group).

## EXERCISE 3

Consider the action of the orthogonal group $O_2$ on $\mathbb{R}^2$. Describe $\text{fix}(\text{ref}_{\frac{\pi}{2}})$ and $\text{fix}(\text{rot}_\pi)$.

# 11.3 Counting Orbits

The next theorem generalizes the observation in the previous section that the subtotal for each orbit is the order of the group.

### • *Theorem 4*

Let $G$ be a finite group acting on a finite set $X$ and let $O$ be an orbit of that action. Then

$$|G| = \sum_{x\in O} |\text{stab}(x)|. \qquad \bullet$$

PROOF

Let $x \in O$. By the finite form of the orbit-stabilizer theorem, $|O| \times |\text{stab}(x)| = |G|$ so $|\text{stab}(x)| = \dfrac{|G|}{|O|}$. Thus

$$\sum_{x\in O} |\text{stab}(x)| = |O| \times \frac{|G|}{|O|} = |G|. \qquad \bullet$$

We can use this result to find formulae for the number of orbits in terms of either stabilizers or fixed sets.

### • *Theorem 5 (The Orbit-Counting Theorem)*

Let $G$ be a finite group acting on a finite set $X$. Then

$$\text{the number of orbits} = \frac{1}{|G|}\sum_{x\in X} |\text{stab}(x)| = \frac{1}{|G|}\sum_{g\in G} |\text{fix}(g)|. \qquad \bullet$$

PROOF

We have seen in Section 12.1 that both these sums are equal to the number of pairs $(g, x)$ where $g * x = x$. Suppose that there are $n$ orbits, $O1, O2, \ldots, On$. Since the orbits partition $X$, the sum $\sum_{x\in G} |\text{stab}(x)|$ is the sum of the subtotals for each orbit. By Theorem 4, each subtotal is $|G|$, that is

$$\sum_{x\in O1} |\text{stab}(x)| + \sum_{x\in O2} |\text{stab}(x)| + \ldots + \sum_{x\in On} |\text{stab}(x)| = |G| + |G| + \ldots + |G| = n \times |G|.$$

Hence

$$\sum_{x \in G} |\operatorname{stab}(x)| = n \times |G|.$$

The result follows on division by $|G|$. ●

In the applications we shall see later, the formula involving fixed sets is more practical than the one involving stabilizers. Although the theorem is often attributed to **Burnside** he may not have been the first to prove it.

### *Example 3*

Let $n$ be an odd number and divide a square into $n^2$ small squares. The group $D_4$ of symmetries of the square acts on these. Let us apply the Orbit-Counting Theorem to find the number of orbits. All $n^2$ squares are fixed by $e$ and each of the rotations $r_1, r_2, r_3$ fixes only the centre square. Each reflection fixes $n$ squares, namely those through which the line of reflection passes. Thus

$$|\operatorname{fix}(e)| = n^2, |\operatorname{fix}(r_1)| = |\operatorname{fix}(r_2)| = |\operatorname{fix}(r_3)| = 1,$$

$$|\operatorname{fix}(s_1)| = |\operatorname{fix}(s_2)| = |\operatorname{fix}(s_3)| = |\operatorname{fix}(s_4)| = n$$

and the number of orbits is

$$\frac{1}{8}(n^2 + 4n + 3) = \frac{1}{8}(n + 1)(n + 3).$$

**EXERCISE 4**

Let $n$ be an even number and divide the square into $n^2$ small squares. Apply the Orbit-Counting Theorem to find the number of orbits for the action of $D_4$ on these $n^2$ squares.

## 11.4 Further Exercises on Chapter 11

**EXERCISE 5**

Let $G$ be the cyclic subgroup of $S_8$ generated by $(1\ 2\ 3)(5\ 6\ 7\ 8)$. Then $G$ acts on the set $\{1, 2, 3, 4, 5, 6, 7, 8\}$ by $g * x = g(x)$.

(i) Find orb(5) and stab(5).

(ii) Display the bijection between orb(5) and the set of left cosets of stab(5) given by the Orbit-Stabilizer Theorem.

## EXERCISE 6

Consider the action of the symmetric group $S_3$ on polynomials in $x_1, x_2$ and $x_3$. Let $p = x_1^2 + x_3^2$.

(i) Find $\operatorname{orb}(p)$ and $\operatorname{stab}(p)$.

(ii) Display the bijection between $\operatorname{orb}(p)$ and the set of left cosets of $\operatorname{stab}(p)$ given by the Orbit-Stabilizer Theorem.

## EXERCISE 7

Consider the action of the group $GL_2(\mathbb{R})$ of all $2 \times 2$ invertible real matrices on $\mathbb{R}^2$ by matrix multiplication. If $P = \begin{pmatrix} 1 \\ 1 \end{pmatrix}$ then $\operatorname{stab}(P) = \left\{ \begin{pmatrix} a & 1-a \\ c & 1-c \end{pmatrix} : a, c \in \mathbb{R} \right\}$.

(i) Show that $\operatorname{orb}(P) = \left\{ \begin{pmatrix} x \\ y \end{pmatrix} : xy \neq 0 \right\}$.

(ii) For each $v \in \operatorname{orb}(P)$, find $\operatorname{send}_P(v)$ and express your answer as a left coset of $\operatorname{stab}(P)$.

## EXERCISE 8

Let $H = \left\{ \begin{pmatrix} \cos\theta & -2\sin\theta \\ \frac{1}{2}\sin\theta & \cos\theta \end{pmatrix} : \theta \in \mathbb{R} \right\}$, a subgroup of $GL_2(\mathbb{R})$. Let $v = \begin{pmatrix} 1 \\ 0 \end{pmatrix}$.

(i) Show that $\operatorname{orb}(v) = \left\{ \begin{pmatrix} \cos\theta \\ \frac{1}{2}\sin\theta \end{pmatrix} : \theta \in \mathbb{R} \right\}$.

(ii) Find $\operatorname{stab}(v)$.

(iii) For each $w \in \operatorname{orb}(v)$, find $\operatorname{send}_v(w)$ and express your answer as a left coset of $\operatorname{stab}(v)$.

## EXERCISE 9

Let $n$ be a positive integer and divide a square into $n^2$ small squares. The group $D_4$ of symmetries of the square acts on the set of $2 \times 2$ squares within the square, of which there are $(n-1)^2$. If $n$ is even, let $m = \frac{n}{2}$ and, if $n$ is odd, let $m = \frac{n-1}{2}$. Show that, in each case, the number of orbits is $\frac{1}{2}m(m+1)$. Investigate the number of orbits for other shapes, for example $1 \times 3$ and $3 \times 1$ rectangles.

## EXERCISE 10

Let $G$ be a group acting on a set $X$. Show that $\operatorname{fix}(g) = \operatorname{fix}(g^{-1})$ for all $g \in G$.

# 12 • Colouring Problems

The aim of this chapter is to apply the Orbit-Counting Theorem from the previous chapter to problems about the number of ways of colouring various geometric patterns.

## 12.1 Colouring Problems

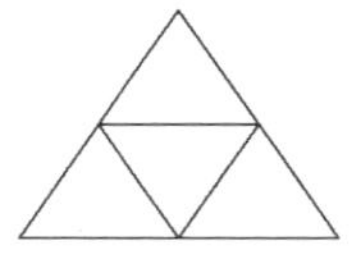

**Fig 12.1** Triangular tile divided into four.

Suppose that we have a tile in the shape of an equilateral triangle which has been divided into four smaller triangles as shown in Fig 12.1. If each of the smaller triangles is coloured (on one side of the tile) in one of red, blue or yellow, how many essentially different colourings are possible?

In order to solve this problem we must first decide what we mean by essentially different colourings. If we colour the top triangle blue and the other three red, this is essentially the same as colouring the bottom right blue and the others red since we can get from one to the other by simply rotating the tile. If, on the other hand, we colour the central triangle blue and the others red we have an essentially different colouring from the first two, since we cannot obtain it by rotating the others.

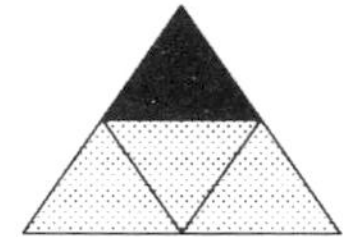
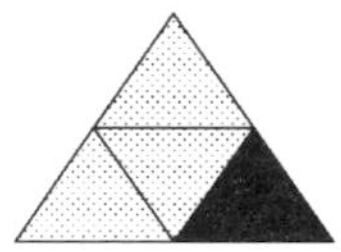
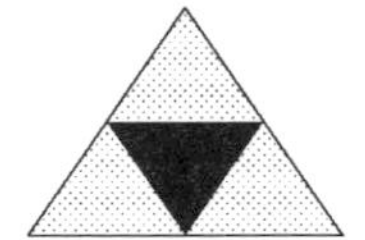

**Fig 12.2** Colourings of a triangular tile.

If one colouring can be obtained from another by a rotation of the tile then it is essentially the same as the first. The rotation group $G = \{e, \mathrm{rot}_{\frac{2\pi}{3}}, \mathrm{rot}_{\frac{4\pi}{3}}\}$ of the triangle acts on the set of colourings. All colourings which are essentially the same belong to the same orbit of the action.

To find the number of essentially different colourings we count the number of orbits using the Orbit-Counting Theorem. The set of possible colourings has $3^4$ elements since there are four triangles which can each be one of three colours. To apply the Orbit-Counting Theorem we need to find $|\,\mathrm{fix}(g)|$ for each of the three elements $g$ of $G$. Now every colouring is fixed by the identity and so $|\,\mathrm{fix}(e)| = 3^4 = 81$. When applying $\mathrm{rot}_{\frac{2\pi}{3}}$ to the

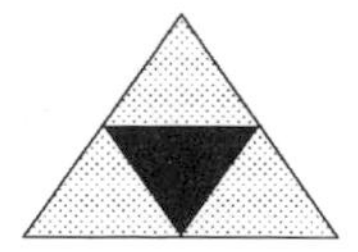

**Fig 12.3** Colouring of a triangular tile fixed by a rotation.

triangle any corner is moved to an adjacent corner. Thus, the colourings fixed by this rotation are those in which all three corner triangles are the same colour. The central triangle can be any colour. There are three choices for the central colour and three for the corner colour so there are $3^2$ colourings fixed by this rotation. The same reasoning applies to the other rotation. Therefore

$$\text{number of orbits} = \frac{1}{|G|}(|\operatorname{fix}(e)| + |\operatorname{fix}(\mathrm{rot}_{\frac{2\pi}{3}})| + |\operatorname{fix}(\mathrm{rot}_{\frac{4\pi}{3}})|) = \frac{81+9+9}{3} = 33.$$

Thus, there are 33 essentially different colourings.

What happens if we now suppose that the tiles are made of a material such as glass which shows the colourings on both sides? The set of all possible colourings is the same but now more are equivalent to each other because we have a bigger group, including reflections, acting on the triangle. Figure 12.4 shows two colourings which can be obtained from each other by reflection but not rotation. The bigger group is the dihedral group $D_3$, the group of all symmetries of the triangle, including reflections as well as rotations.

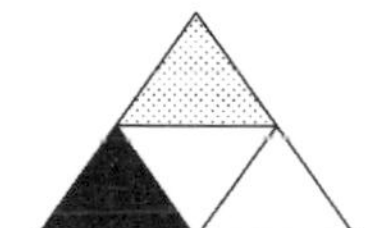

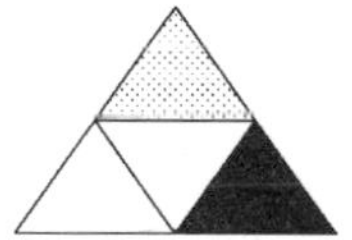

**Fig 12.4** Equivalent colourings for a glass tile.

How many colourings are fixed by one of the reflections? As can be seen from Fig 12.5,

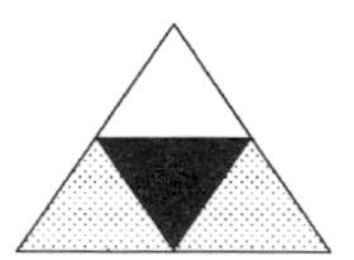

**Fig 12.5** Colourings of a triangular tile fixed by a reflection.

two corner triangles must be the same but the third corner and the centre triangle can both be different. There are three choices of colour for the centre, three for the pair and three for the third corner. Hence, there are $3^3 = 27$ colourings fixed by a reflection. This applies to all three reflections. The group $D_3$ has order 6 so

$$\text{number of orbits} = \frac{81+9+9+27+27+27}{6} = 30.$$

Thus, there are 30 essentially different colourings in this case.

### Example I

In how many essentially different ways can the square glass tile shown below, which can be turned over, be coloured

(i) using $n$ colours;

(ii) so that there are six purple regions and two yellow regions?

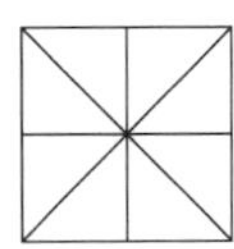

The group acting here is $D_4$, the group of symmetries of the square.

(i) There are $n$ choices of colour for each of the eight regions so there are $n^8$ colourings. All are fixed by $e$ so $|\operatorname{fix}(e)| = n^8$.

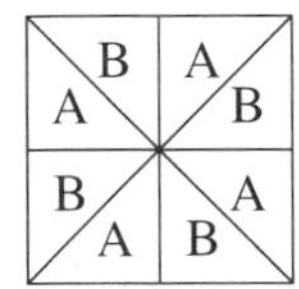


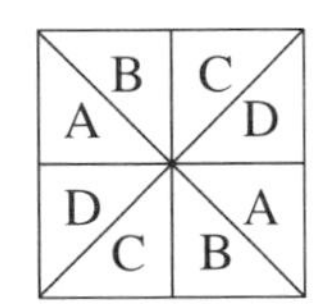


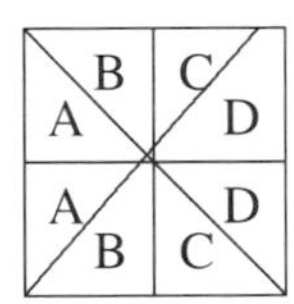


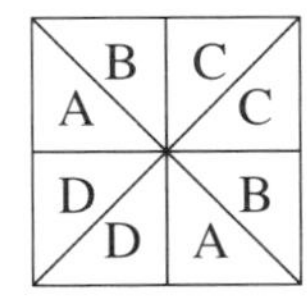


**Fig 12.6** Fixed colourings for $r_1, r_2, s_1, s_2$.

The colourings fixed by $r_1$ must have the same colour in each of the regions marked A in the first diagram in Fig 12.6 and a second colour in each of the regions marked B. Hence, $|\operatorname{fix}(r_1)| = n^2$. The other three diagrams indicate that $|\operatorname{fix}(r_2)| = n^4 = |\operatorname{fix}(s_1)| = |\operatorname{fix}(s_2)|$. The colourings fixed by $r_3$ are the same as those fixed by its inverse $r_1$ and the diagrams for the reflections $s_3$ and $s_4$ are similar to those for $s_1$ and $s_2$. Hence, $|\operatorname{fix}(r_3)| = n^2$ and $|\operatorname{fix}(s_3)| = n^4 = |\operatorname{fix}(s_4)|$. By the Orbit-Counting Theorem, the total number of essentially different colourings is

$$\frac{1}{8}(n^8 + 5n^4 + 2n^2).$$

(ii) The total number of colourings is ${}_8C_2 = 28$ so $|\operatorname{fix}(e)| = 28$.

There are no colourings fixed by $r_1$ because we cannot have only two yellow regions, see Fig 12.6. Thus, $|\operatorname{fix}(r_1)| = 0$ and similarly $|\operatorname{fix}(r_3)| = 0$.

In a colouring fixed by $r_2$, any one of four pairs can be yellow so $|\operatorname{fix}(r_2)| = 4$. Similarly $|\operatorname{fix}(s_1)| = |\operatorname{fix}(s_2)| = |\operatorname{fix}(s_3)| = |\operatorname{fix}(s_4)| = 4$.

By the Orbit-Counting Theorem, the total number of essentially different colourings is

$$\frac{1}{8}(28 + 4 + 16) = 6.$$

## EXERCISE 1

In (ii) of the above example, how many essentially different colourings are there if the tile cannot be turned over (for example, if it is a ceramic floor tile)? Find two colourings which can be regarded as the same when we allow reflections but are different when we only allow rotations.

## EXERCISE 2

(i) In how many essentially different ways can the rectangular glass tile shown below, which can be turned over, be coloured

(a) using $n$ colours;

(b) with two blue regions and two red regions?

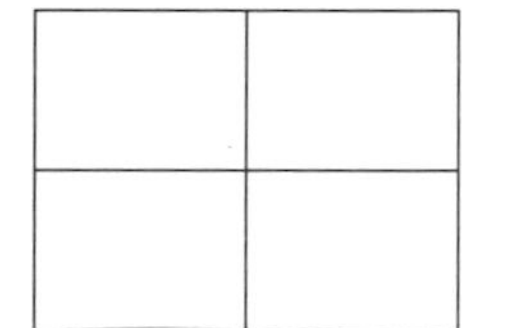

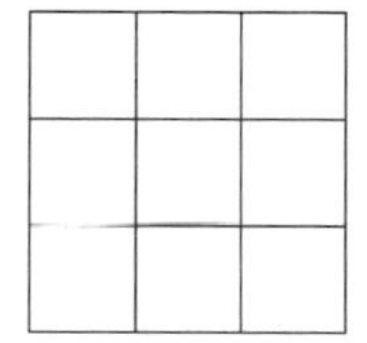

(ii) In how many ways can a square divided into nine smaller squares be coloured using two colours when

(a) the tile is coloured on one side;

(b) the tile is made of glass and can be turned over?

# 12.2 Groups of Symmetries in Three Dimensions

The idea of the group of symmetries of a plane geometric figure, such as a square, extends to three-dimensional objects, such as a cube or regular tetrahedron. In this section, we describe the groups of symmetries of these two figures. The description will be much easier to follow if you have models of the solids described available. Models can be made from paper or card. There are now at least two manufacturers of plastic shapes that can be fitted together to make excellent models.

## The cube

For a rotation in three dimensions, we need to specify an axis of rotation and an angle. A cube has three different types of axis of rotation. There are three axes through the centres of opposite faces. About each of these, there are rotations through 0, $\frac{\pi}{2}$, $\pi$ and $\frac{3\pi}{2}$. There are four axes through opposite vertices. About each of these, there are rotations through 0, $\frac{2\pi}{3}$ and $\frac{4\pi}{3}$. Finally, there are six axes through midpoints of opposite edges. About each of these, there are rotations through 0 and $\pi$. In all, there are 24 rotations. These are listed in Table 12.1 and they form the **rotation group** of the cube.

The rotation group $G$ of the cube acts on the eight vertices, the six faces, the 12 edges and the four diagonals of the cube. These actions give rise to homomorphisms from $G$ to $S_8$, $S_6$, $S_{12}$ and $S_4$. Those for the faces and diagonals are shown in Table 12.1. The final column indicates that all 24 permutations of the diagonals occur. Hence, the homomorphism $f : G \to S_4$ arising from the action on the diagonals is an isomorphism. Thus, $G \cong S_4$.

| Axis type | Angle of rotation | Number of axes | Number of rotations | Permutation of faces | Permutation of diagonals |
|---|---|---|---|---|---|
| face–face | $\frac{\pi}{2}$ or $\frac{3\pi}{2}$ | 3 | 6 | $(a\ b\ c\ d)(e)(f)$ | $(a\ b\ c\ d)$ |
| face–face | $\pi$ | 3 | 3 | $(a\ c)(b\ d)(e)(f)$ | $(a\ c)(b\ d)$ |
| vertex–vertex | $\frac{2\pi}{3}$ or $\frac{4\pi}{3}$ | 4 | 8 | $(a\ b\ c)(d\ e\ f)$ | $(a\ b\ c)(d)$ |
| edge–edge | $\pi$ | 6 | 6 | $(a\ b)(c\ d)(e\ f)$ | $(a\ b)(c)(d)$ |
| any | 0 | 16 | 1 | id | id |

**Table 12.1** Rotations of the cube.

To specify a reflection in three dimensions, we need a plane of symmetry. The cube has nine planes of symmetry. Three of these are parallel to faces and six pass through opposite edges. Not all symmetries of the cube are rotations and reflections. The group of symmetries has order 48 and acts on the six faces of the cube. The orbit of the top face has order 6 because it can be sent to any other face by an appropriate rotation. Each element of the stabilizer $S$ of the top face sends that face to itself and so performs a symmetry of the square on that face. This gives an isomorphism between $S$ and the group of symmetries of the square. Thus, $S$ has order 8. In accordance with the finite form of the Orbit-Stabilizer Theorem, the group of symmetries has order $6 \times 8 = 48$. Only 33 of the 48 symmetries are rotations and reflections. The remaining 15 symmetries can each be written in the form $sr$ where $s$ is a reflection and $r$ is a rotation.

## The tetrahedron

A regular tetrahedron has four faces each of which is an equilateral triangle. The **rotation group** of the regular tetrahedron has order 12. There are four axes of rotation through a vertex and the middle of the opposite face, and six through the midpoints of opposite edges. These are shown in Table 12.2 together with the corresponding permutations of the four vertices.

| Axis type | Angle of rotation | Number of axes | Number of rotations | Permutation of vertices |
|---|---|---|---|---|
| vertex–face | $\frac{2\pi}{3}$ or $\frac{4\pi}{3}$ | 4 | 8 | $(a\ b\ c)(d)$ |
| edge–edge | $\pi$ | 3 | 3 | $(a\ b)(c\ d)$ |
| any | 0 | 7 | 1 | id |

**Table 12.2** Rotations of the tetrahedron.

The final column indicates that all 12 even permutations in $S_4$ occur and hence that the rotation group is isomorphic to the alternating group $A_4$.

The group of symmetries of the tetrahedron has order 24 and is isomorphic to $S_4$. When it acts on the vertices, the orbit of any face has four elements and the stabilizer of any face has order 6. There are six reflections.

**EXERCISE 3**

A regular octahedron has eight faces all of which are equilateral triangles. What are the orders of its group of symmetries and its rotation group?

**EXERCISE 4**

A rectangular prism has six rectangular faces. It has 12 edges, four of each of three lengths $a, b, c$. When $a = b = c$, it is a cube. Find the orders of its group of symmetries and its rotation group

(i) when $a, b, c$ are distinct;

(ii) when $a = b \neq c$.

## 12.3 Three-dimensional Colouring Problems

### *Example 2*

In how many essentially different ways can the faces of a regular tetrahedron be coloured using $n$ colours, each face being a single colour.

Because only rotations can be physically performed, we use the action of the rotation group of the tetrahedron rather than the full group of symmetries. Table 12.3 shows the sizes of the fixed sets for the various rotations. For an axis through opposite edges, each

| Axis type | Angle | $\lvert\text{fix}(g)\rvert$ | Number of rotations | Total |
|---|---|---|---|---|
| edge–edge | $\pi$ | $n^2$ | 3 | $3n^2$ |
| vertex–face | $\frac{2\pi}{3}$ or $\frac{4\pi}{3}$ | $n^2$ | 8 | $8n^2$ |
| any | 0 | $n^4$ | 1 | $n^4$ |

**Table 12.3** Colourings fixed under symmetries of the tetrahedron.

pair of faces meeting at one edge must be the same colour if the colouring is fixed by the rotation. For an axis through a vertex, the three faces surrounding the vertex must be the same colour and the base triangle can be different. Although there are two non-identity rotations about this type of axis, they both fix the same colourings.

The number of essentially different colourings is therefore

$$\frac{3n^2 + 8n^2 + n^4}{12} = \frac{n^2(n^2 + 11)}{12}.$$

### Example 3

In how many essentially different ways can we colour the faces of a cube using red and blue?

A cube has six faces so there are $2^6 = 64$ possible colourings. The numbers of colourings fixed by each type of rotation are shown in Table 12.4. It would be helpful to have a cube available for inspection when reading this table.

| Type of axis | Angle | Number of rotations | $\lvert \text{fix}(g) \rvert$ | Total number fixed |
|---|---|---|---|---|
| face–face | $\frac{\pi}{2}$ or $\frac{3\pi}{2}$ | 6 | $2^3$ | 48 |
| face–face | $\pi$ | 3 | $2^4$ | 48 |
| vertex–vertex | $\frac{2\pi}{3}$ or $\frac{4\pi}{3}$ | 8 | $2^2$ | 32 |
| edge–edge | $\pi$ | 6 | $2^3$ | 48 |
| any | 0 | 1 | $2^6$ | 64 |

**Table 12.4** Table of colourings fixed by rotations of the cube.

The number of essentially different colourings is $\frac{1}{24}(48+48+32+48+64) = \frac{240}{24} = 10.$

## EXERCISE 5

In how many essentially different ways can the faces of a cube be coloured with four red faces and two blue faces?

## EXERCISE 6

In how many essentially different ways can a tetrahedron be coloured with two green faces and two yellow faces?

# 12.4 Further Exercises on Chapter 12

## EXERCISE 7

In how many essentially different ways can the equilateral triangular tile below, which can be turned over, be coloured

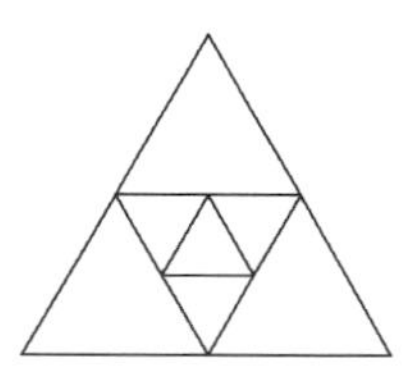

(i) using $n$ colours;

(ii) so that there are four blue regions and three green regions?

## EXERCISE 8

A square glass tile, which can be turned over, is to be made from eight triangular pieces of coloured glass in the pattern below.

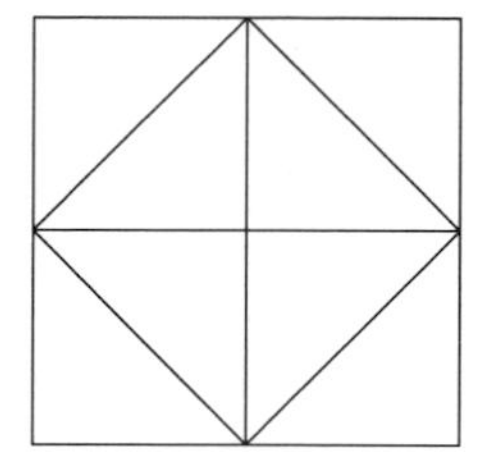

Find the number of essentially different ways in which this can be done

(i) using $n$ colours;

(ii) using four red pieces and four green pieces;

(iii) using two turquoise pieces and $n$ colours, not including turquoise, for the remaining six pieces.

## EXERCISE 9

In how many essentially different ways can the front face of the hexagonal tile below be coloured using red and blue?

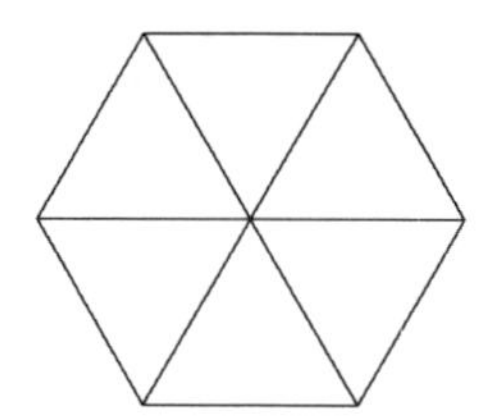

(Note: the reference to the front face indicates that the tile cannot be turned over, so the group whose action should be considered here is the rotation group of the hexagon.)

## EXERCISE 10

In the well-known game of noughts and crosses, a square $3 \times 3$ grid is filled in with crosses and noughts. In how many essentially different ways can this be done with five crosses and four noughts?

## EXERCISE 11

In how many essentially different ways can a tetrahedron be coloured

(i) with one blue face and three red faces;

(ii) with one blue face, one red face, one green face, and one yellow face?

## EXERCISE 12

Consider the isomorphism $f : G \to S_8$ arising from the action of the rotation group $G$ of the cube on the vertices of the cube. Find the cycle type of the permutation $f(r)$ when

(i) $r$ is rotation through $\frac{\pi}{2}$ about an axis through midpoints of opposite faces;

(ii) $r$ is rotation through $\frac{2\pi}{3}$ about an axis through opposite vertices;

(iii) $r$ is rotation through $\pi$ about an axis through opposite edges.

## EXERCISE 13

(i) Describe the rotations of a regular octahedron and show that its rotation group is isomorphic to $S_4$.

(ii) A Christmas tree is to be decorated with regular octahedra with faces painted in silver and gold. Suppose that we would like a complete set of all essentially different colourings of octahedra in silver and gold. How many do we need?

## EXERCISE 14

An icosahedron has 20 faces, each of which is an equilateral triangle. A dodecahedron has 12 faces, each of which is a regular pentagon. Investigate their groups of symmetries and their rotation groups.

# 13 • Conjugates, Centralizers and Centres

In several of the groups we have met there are instances where different elements are in some sense similar. For example, in the symmetric group $S_4$ there are eight 3-cycles which can be regarded as being similar. In the orthogonal group $O_2$ it is reasonable to regard all the reflections as being similar. This is not true for the rotations in $O_2$ because these can have any order, finite or infinite. However $\text{rot}_{\frac{2\pi}{3}}$ and $\text{rot}_{\frac{4\pi}{3}}$, both of order 3, are similar. Indeed they are both rotations through $\frac{2\pi}{3}$, one anticlockwise and the other clockwise. One aim of this chapter is to formalize this notion of similarity. The ideas involved in this have some striking consequences for group theory in general.

## 13.1 Conjugates

Let $G$ be a group and let $a, g \in G$. The **conjugate** of $a$ by $g$ is the element $gag^{-1}$ of $G$. We can compute conjugates in the orthogonal group $O_2$ using Table 1.1

$$\text{ref}_\phi \, \text{rot}_\theta (\text{ref}_\phi)^{-1} = \text{ref}_{\phi-\theta} \, \text{ref}_\phi = \text{rot}_{-\theta} \text{ and } \text{rot}_\phi \, \text{ref}_\theta (\text{rot}_\phi)^{-1} = \text{ref}_{\phi+\theta} \, \text{rot}_{-\phi} = \text{ref}_{2\phi+\theta} \, .$$

Let $G$ be a group and let $g \in G$. Consider the function $f : G \to G$ which sends each element $a$ to its conjugate by $g$, $f : a \mapsto gag^{-1}$. Let $a, b \in G$. Then

$$f(a)f(b) = gag^{-1}gbg^{-1} = gabg^{-1} = f(ab).$$

Thus, $f$ is a homomorphism. It is easy to check that the function $a \mapsto g^{-1}ag$, the conjugate of $a$ by $g^{-1}$, is an inverse for $f$ and hence that $f$ is an isomorphism. By Theorem 3 of Chapter 9, we can see that $a$ and $gag^{-1}$ have the same order.

### EXERCISE 1

Compute the following conjugates

(i) $s_3r_2(s_3)^{-1}$ and $r_1s_1(r_1)^{-1}$ in $D_4$;

(ii) $\text{ref}_\phi \, \text{ref}_\theta (\text{ref}_\phi)^{-1}$ and $\text{rot}_\phi \, \text{rot}_\theta (\text{rot}_\phi)^{-1}$ in $O_2$.

## 13.2 Conjugacy Classes

It appears that if $b$ is a conjugate of $a$ then $a$ and $b$ have similar properties. This suggests that conjugacy gives rise to an equivalence relation. We saw in Chapter 8 how a group action determines an equivalence relation in which the equivalence classes are the orbits. So is there a group action here? The answer is that the group $G$ is acting on *itself*.

## ● Theorem 1

Let $G$ be a group. Then $G$ acts on $G$ by the rule

$g * a = gag^{-1}$ for all $a, g \in G$. ●

PROOF

**GA1:** $e * a = eae^{-1} = a$ for every $a \in G$.

**GA2:** $(gh) * a = gha(gh)^{-1} = ghah^{-1}g^{-1} = g * (hah^{-1}) = g * (h * a)$ for all $g \in G, h \in G$ and all $a \in G$. ●

It now follows from Section 8.4 that there is an equivalence relation $R$ on $G$ given by the rule

$aRb$ if and only if $b = gag^{-1}$ for some $g \in G$.

The equivalence class of $a$ for this relation is the orbit of $a$ for this action and consists of all the conjugates $gag^{-1}$ of $a$ in $G$. It is written $\text{conj}_G\, a$ and is called the **conjugacy class** of $a$ in $G$. Thus

$$\text{conj}_G\, a = \{gag^{-1} : g \in G\}.$$

## Example 1

The conjugacy class in $O_2$ of any rotation $\text{rot}_\theta$ consists of $\text{rot}_\theta$, which is the conjugate of $\text{rot}_\theta$ by any rotation, and $\text{rot}_{-\theta}$, which is the conjugate of $\text{rot}_\theta$ by any reflection. The conjugacy class of any reflection $\text{ref}_\theta$ consists of all reflections since $\text{ref}_\phi = \text{rot}_\psi \text{ref}_\theta (\text{rot}_\psi)^{-1}$ where $\psi = \frac{\phi-\theta}{2}$.

## Example 2

The symmetric group $S_3$ has three transpositions, $(1\ 2)$, $(1\ 3)$ and $(2\ 3)$, and two 3-cycles, $(1\ 2\ 3)$ and $(1\ 3\ 2)$. If you calculate the conjugate $\theta(1\ 2)\theta^{-1}$ for each $\theta \in S_3$ you should obtain the three transpositions, each occurring twice. The conjugates of $(1\ 2\ 3)$ are the two 3-cycles and the only conjugate of the identity permutation is id itself. The conjugacy classes in $S_3$ are

$$\{(1\ 2), (1\ 3), (2\ 3)\},\ \{(1\ 2\ 3), (1\ 3\ 2)\},\ \text{and } \{\text{id}\}.$$

Example 2 illustrates the general property that conjugacy classes always partition a group. It also illustrates the following list of basic properties of conjugacy classes. Most of these are deduced from those of equivalence classes in general.

- For each element $a$ of $G$, $a = eae^{-1} \in \text{conj}_G\, a$.
- For $a, b \in G$, $\text{conj}_G\, a = \text{conj}_G\, b$ if and only if $b$ is a conjugate of $a$.
- $\text{conj}_G\, e = \{e\}$ for any group $G$ (because $geg^{-1} = e$ for all $g \in G$).
- All elements of a conjugacy class have the same order.

When the group $G$ is finite the total number of elements in $G$ must be the sum of the numbers of elements in each conjugacy class. This gives the following theorem.

**Theorem 2**

Let $G$ be a finite group and let $C_1, C_2, \ldots, C_k$ be its distinct conjugacy classes. Then

$$|G| = |C_1| + |C_2| + \ldots + |C_k|.$$

The equation in this theorem is usually called the **class equation** for $G$. The class equation for $S_3$ is simply 6=1+3+2.

We have seen that, in the orthogonal group $O_2$, any two reflections are conjugate. However, as the next exercise shows, in $D_4$, the idea of conjugacy distinguishes between the two reflections $s_2$ and $s_4$ in the diagonals, which form one class, and the other two reflections, $s_1$ and $s_3$. The elements $s_1 = \text{ref}_0$ and $s_2 = \text{ref}_{\frac{\pi}{2}}$, both of order 2, are not conjugate in $D_4$.

**EXERCISE 2**

Show that the conjugacy classes of $D_4$ are

$$\{e\}, \{\text{rot}_\pi\}, \{\text{rot}_{\frac{\pi}{2}}, \text{rot}_{\frac{3\pi}{2}}\}, \{\text{ref}_0, \text{ref}_\pi\} \text{ and } \{\text{ref}_{\frac{\pi}{2}}, \text{ref}_{\frac{3\pi}{2}}\}.$$

Write down the class equation for $D_4$.

**EXERCISE 3**

For which angles $\alpha$, $0 \leqslant \alpha < 2\pi$, does $\text{rot}_\alpha$ have only one conjugate in $O_2$? (In other words, when is $\text{rot}_{-\alpha} = \text{rot}_\alpha$?)

## 13.3 Conjugacy Classes in $S_n$

In our discussion of the conjugacy classes of $S_3$, all permutations with similar cycle decompositions, for example all the transpositions, were in the same conjugacy class. We shall now see that this is true for $S_n$ in general. In order to do so we need some notation to explain what we mean by 'similar' in this context.

In what follows, we include all cycles of length one in cycle decompositions. Any element $\alpha$ of $S_n$ has a cycle decomposition, $\alpha = \alpha_1\alpha_2 \ldots \alpha_s$, as the product of $s$ disjoint cycles. If each $\alpha_i$ has length $k_i$ then $k_1 + k_2 + \ldots + k_s = n$. Disjoint cycles commute so we can assume that $k_1 \geqslant k_2 \geqslant \ldots \geqslant k_s$, in other words we arrange the cycles in descending order of length. We then say that $\alpha$ has **cycle type** $(k_1, k_2, \ldots, k_s)$. For example, in $S_9$, $\alpha = (1\ 5\ 3)(2\ 9)(4)(6\ 8\ 7) = (1\ 5\ 3)(6\ 8\ 7)(2\ 9)(4)$ has cycle type (3, 3, 2, 1). For any element $\alpha$ of $S_n$, the sum of the numbers in its cycle type must be $n$. In $S_n$, id has cycle type $(1, 1, \ldots, 1)$ and, at the other extreme, an $n$-cycle has cycle type $(n)$.

**EXERCISE 4**

Write down the cycle types of each of the following permutations in $S_9$.

(i) (1 2 4)(3 5)(6 9 8 7);

(ii) (1 2 3 5)(6 7 8 9);

(iii) (1 9).

Now consider the two permutations $\alpha = (1\ 2\ 3\ 4)(5\ 6\ 7)(8\ 9)$ and
$\theta = \begin{pmatrix} 1 & 2 & 3 & 4 & 5 & 6 & 7 & 8 & 9 \\ 4 & 6 & 9 & 5 & 3 & 1 & 7 & 2 & 8 \end{pmatrix}$. As an exercise, you should verify that
$\theta\alpha\theta^{-1} = (4\ 6\ 9\ 5)(3\ 1\ 7)(2\ 8)$. Notice that this permutation has the same cycle type as $\alpha$ and that it is obtained from $\alpha$ by applying $\theta$ to each of the numbers appearing in each cycle of $\alpha$. This is typical of what happens in general.

## • *Theorem 3*

Let $\alpha, \theta \in S_n$. The conjugate $\theta\alpha\theta^{-1}$ has the same cycle type as $\alpha$ and is obtained from $\alpha$ by applying $\theta$ to each of the numbers appearing in each cycle of $\alpha$. •

PROOF

Consider first the case where $\alpha = (a_1\ a_2 \ldots a_k)$ is a cycle of length $k$. Let $b_1 = \theta(a_1),\ b_2 = \theta(a_2), \ldots, b_k = \theta(a_k)$ and let $\beta$ be the cycle $(b_1\ b_2\ \ldots\ b_k)$. We want to show that $\theta\alpha\theta^{-1} = \beta$. To show that two permutations in $S_n$ are equal we need to show that they have the same effect on each number $x$ in $\{1, 2, \ldots, n\}$. If $x$ appears in $\beta$ then we can rewrite the cycles, without changing their cyclic order, so that $x = b_1$. Then

$$\theta\alpha\theta^{-1}(b_1) = \theta\alpha(a_1) = \theta(a_2) = b_2 = \beta(b_1).$$

Thus, $\theta\alpha\theta^{-1}(x) = \beta(x)$ for all $x$ appearing in $\beta$. On the other hand, if $x$ does not appear in $\beta$ then $\beta(x) = x$ and, because $\theta^{-1}(x)$ does not appear in $\alpha$, $\alpha\theta^{-1}(x) = \theta^{-1}(x)$. Hence

$$\theta\alpha\theta^{-1}(x) = \theta\theta^{-1}(x) = x = \beta(x).$$

We have now shown that $\beta(x) = \theta\alpha\theta^{-1}(x)$ for all $x$ and so $\beta = \theta\alpha\theta^{-1}$. This establishes the formula

$$\theta(a_1\ a_2\ \ldots a_k)\theta^{-1} = (\theta(a_1)\ \theta(a_2) \ldots \theta(a_k)).$$

Now consider the general case where $\alpha$ has cycle decomposition $\alpha_1\alpha_2 \ldots \alpha_s$. For each $i$, let $\beta_i$ be the cycle obtained by applying $\theta$ to each number appearing in $\alpha_i$. By the above formula, each $\theta\alpha_i\theta^{-1} = \beta_i$. Hence

$$\theta\alpha\theta^{-1} = \theta(\alpha_1\alpha_2 \ldots \alpha_s)\theta^{-1} = \theta\alpha_1\theta^{-1}\theta\alpha_2\theta^{-1}\theta \ldots \theta^{-1}\theta\alpha_s\theta^{-1} = \beta_1\beta_2 \ldots \beta_s$$

has the same cycle type as $\alpha$. •

Given any two permutations $\alpha$ and $\beta$ in $S_n$ of the same cycle type, there is a permutation $\theta \in S_n$ such that $\beta$ can be obtained from $\alpha$ by applying $\theta$ to each number appearing in $\alpha$. By Theorem 3, $\beta = \theta\alpha\theta^{-1}$. For example, if $\alpha = (1\ 2\ 3)(4\ 5)$ and $\beta = (1\ 6\ 2)(3\ 4)$ in $S_6$ then we take $\theta = \begin{pmatrix} 1 & 2 & 3 & 4 & 5 & 6 \\ 1 & 6 & 2 & 3 & 4 & 5 \end{pmatrix}$. Thus, any two permutations in $S_n$ of the same cycle type are conjugates in $S_n$. Combining this with Theorem 3, we see that each conjugacy class in $S_n$ consists of all permutations with a given cycle type.

**EXERCISE 5**

Let $\alpha = (1\ 7)(6\ 5\ 4\ 3)(2\ 9\ 8)$, $\beta = (1\ 2\ 3\ 6)(4\ 5\ 8)(7\ 9)$. Write down $\theta \in S_9$ such that $\beta = \theta\alpha\theta^{-1}$.

**EXERCISE 6**

Write down two elements of $S_4$ which have the same order but are not conjugates in $S_4$.

## 13.4 Centralizers

We have seen that conjugation is a group action and have considered its orbits, the conjugacy classes. Here we consider the stabilizers of this action. For $a \in G$, the stabilizer of $a$ consists of those elements $g$ of $G$ such that $gag^{-1} = a$. Now if $gag^{-1} = a$ then, multiplying by $g$ on the right, $ga = ga$. Conversely, if $ga = ag$ then, multiplying by $g^{-1}$ on the right, $gag^{-1} = a$. So $gag^{-1} = a$ if and only if $ga = ag$.

The stabilizer of $a$ in $G$ for conjugation is called the **centralizer** of $a$ in $G$ and is written $\text{cent}_G(a)$. Thus, $\text{cent}_G(a) = \{g \in G : ga = ag\}$ consists of all the elements of $G$ which commute with $a$.

Being a stabilizer for a group action, $\text{cent}_G(a)$ is always a subgroup of $G$. In $D_4$ the centralizer of $r_1 = \text{rot}_{\frac{\pi}{2}}$ can be read off from the Cayley table of $D_4$, Table 1.2, from which it is clear that $r_1$ commutes with $e, r_1, r_2$ and $r_3$ but not with any of the reflections. Thus, $\text{cent}_{D_4}(r_1) = \{e, r_1, r_2, r_3\} =< r_1 >$, the cyclic subgroup generated by $r_1$.

It is always true that $< a >\subseteq \text{cent}_G(a)$, because $aa^i = a^{i+1} = a^i a$ for all integers $i$, but in the next exercise there are examples where $< a >\neq \text{cent}_G(a)$.

**EXERCISE 7**

Find the following centralizers. In each case, state whether $\text{cent}_G(a) =< a >$.

(i) $\text{cent}_{S_3}((1\ 2\ 3))$ and $\text{cent}_{S_3}((1\ 2))$.

(ii) $\text{cent}_{D_4}(r_2)$ and $\text{cent}_{D_4}(s_1)$.

**EXERCISE 8**

If $G$ is an abelian group, what can you say about the centralizer of any element of $G$? What is the centralizer of the neutral element $e_G$ in any group $G$?

## 13.5 Centres

Sometimes a centralizer $\text{cent}_G(a)$ is the whole group $G$. In other words $a$ commutes with every element of $G$. Such an element is said to be **central** in $G$. In any group $G$, the set of all central elements in $G$ is called the **centre** of $G$ and is written $Z(G)$. Thus

$$Z(G) = \{a \in G : ga = ag \text{ for all } g \in G\}.$$

An element $a$ of $G$ is central in $G$ if and only if $a \in \text{cent}_G(g)$ for all $g \in G$. In other words, $Z(G)$ is the intersection of all the centralizers $\text{cent}_G(g)$. From this description, it is possible to deduce that $Z(G)$ is always a subgroup but let us check this directly from the subgroup criterion.

**SG1:** For all $g$ in $G$, $ge = eg$ so $e \in Z(G)$ and $Z(G) \neq \emptyset$.

**SG2:** Let $a, b \in Z(G)$. Thus, for all $g \in G$, $ga = ga$ and $gb = bg$. Let $g \in G$. Then

$$g(ab) = (ga)b = (ag)b = a(gb) = a(bg) = (ab)g.$$

Thus, $g(ab) = (ab)g$ for all $g \in G$, that is $ab \in Z(G)$.

**SG3:** Let $a \in Z(G)$ and let $g \in G$. Multiplying throughout the equation $ga = ag$ by $a^{-1}$ on both right and left, we obtain $a^{-1}gaa^{-1} = a^{-1}aga^{-1}$, that is $a^{-1}g = ga^{-1}$. Thus, $a^{-1} \in Z(G)$.

By the subgroup criterion, $Z(G)$ is a subgroup of $G$.

Let $G$ be a group and let $a \in G$. The following are equivalent ways of saying that $a$ is central in $G$.

- $ga = ag$ for all $g \in G$.
- $\text{cent}_G(a) = G$.
- $a \in \text{cent}_G\, g$ for all $g \in G$.
- $gag^{-1} = a$ for all $g \in G$.
- $\text{conj}_G\, a = \{a\}$.

Having many conjugates in $G$ is, loosely speaking, the opposite of commuting with many elements of $G$. In the next section we shall make this comment precise for finite groups.

The centre of an abelian group is the whole group. On the other hand, the centre of a group may be the trivial subgroup, as can be seen from the next exercise. The order of the centre of a group, relative to the order of the whole group, is a measure of how close the group is to being abelian.

**EXERCISE 9**

Using the Cayley table, Table 9.1, check that the only element of $S_3$ which commutes with both (1 2) and (1 2 3) is id and hence that $Z(S_3) = \{\text{id}\}$.

## 13.6 Conjugates and Centralizers

In Chapter 11, the Orbit-Stabilizer Theorem gave a connection between orbits and stabilizers. Applying this to conjugation, we should get a connection between conjugacy classes (the orbits) and centralizers (the stabilizers). In particular, when $G$ is finite, the finite form of the Orbit-Stabilizer Theorem

$$|\text{orb}_G(a)| \times |\text{stab}_G(a)| = |G|,$$

when applied to conjugation, gives the formula in the next theorem.

## Theorem 4

For any element $a$ of any finite group $G$,

$$|\operatorname{conj}_G(a)| \times |\operatorname{cent}_G(a)| = |G|.$$

Thus, $|\operatorname{conj}_G a|$ is always a factor of $|G|$ and is inversely proportional to $|\operatorname{cent}_G a|$. In a group where we have already calculated conjugacy classes, such as $S_n$, this enables us to calculate quickly centralizers and centres.

### Example 3

If $n \geqslant 3$, the conjugacy class in $S_n$ of $(1\ 2\ \ldots\ n)$ has $(n-1)!$ elements, this being the number of different $n$-cycles in $S_n$. Hence, its centralizer has order $n = \frac{n!}{(n-1)!}$ and must be the cyclic subgroup $< (1\ 2\ \ldots\ n) >$. Now let $g = (1\ 2\ \ldots n-1)$. The conjugates of $g$ in $S_n$ are the cycles $h$ of length $n-1$. How many of these are there? There are $n$ choices for the number left out of $h$ and, for each of these, there are $(n-2)!$ different $(n-1)$-cycles. Thus, $g$ has $n \times (n-2)!$ conjugates and so its centralizer has order $n-1$. Any element in the centre of $S_n$ must commute with both $f$ and $g$ and so must be in $\operatorname{cent}(f) \cap \operatorname{cent}(g)$. As $n$ and $n-1$ have highest common factor 1, it follows from Theorem 9(i) of Chapter 10 that $\operatorname{cent}(f) \cap \operatorname{cent}(g) = \{\mathrm{id}\}$. Consequently, the centre of $S_n$ is the trivial subgroup.

Theorem 4 and the class equation have the following striking theoretical consequence. We have just seen that familiar groups may have trivial centres. However, sometimes the nature of the order of the group can preclude this possibility.

## Theorem 5

Let $G$ be a group of prime power order (that is $|G| = p^n$ for some prime number $p$ and positive integer $n$). Then $G$ has a non-trivial centre.

PROOF

Let $C_1, C_2, \ldots, C_k$ be a list of the different conjugacy classes of $G$, numbered so that $C_1 = \operatorname{conj}_G(e) = \{e\}$. Thus, $|C_1| = 1$. We aim to show that there is at least one other conjugacy class with a single element $a$, say, $a \neq e$. For this element $a$ and any element $g$ of $G$ we then have $gag^{-1} = a$ so that $ga = ag$ and $a$ is central.

The class equation says that

$$p^n = 1 + |C_2| + |C_3| + \ldots |C_k|.$$

By Theorem 4, each $|C_i|$ is a factor of $|G| = p^n$. As $p$ is prime, the only factors of $p^n$ are $1, p, p^2, \ldots, p^{n-1}$ and $p^n$. It is impossible for $|C_2|,\ |C_3|, \ldots, |C_k|$ all to be multiples of $p$ because, if they were, then 1 would be a multiple of $p$. Hence, $|C_i| = 1$ for some $i > 1$. As indicated above, if $C_i = \{a\}$ then $a \neq e$ and $a$ is central. Thus, $Z(G) \neq \{e\}$.

This theorem is illustrated by the group $D_4$ which has prime power order $8 = 2^3$. The theorem tells us that, in contrast to $S_4$, the dihedral group $D_4$ must have a non-trivial centre. The element $r_2 = \operatorname{rot}_\pi$ is central. For $D_4$, the class equation is

$$8 = 1 + 1 + 2 + 2 + 2.$$

The idea behind the proof of the theorem is illustrated by the fact that it is impossible to express 8 as a sum of 1s, 2s, 4s and 8s with exactly one 1.

## 13.7 Further Exercises on Chapter 13

### EXERCISE 10

In $S_7$, let $\alpha = (1\ 2\ 3\ 4\ 5\ 6\ 7)$. Find $\theta \in S_7$ of order 3 with $\theta\alpha\theta^{-1} = \alpha^2$.

### EXERCISE 11

How many different conjugacy classes are there in $S_4$? Write down the class equation for $S_4$.

### EXERCISE 12

Find the order of the conjugacy class and the centralizer of each of the following elements of $S_4$

(i) $(1\ 2)(3\ 4)$; (ii) $(1\ 2)$; (iii) $(1\ 2\ 3)$; (iv)$(1\ 2\ 3\ 4)$.

### EXERCISE 13

Let $n > 3$ and, in $S_n$, let $h = (1\ 2 \ldots n-2)$. Find $|\operatorname{cent}_{S_n}(h)|$ and show that $\operatorname{cent}_{S_n}(h) \neq < h >$.

### EXERCISE 14

Investigate the conjugacy classes in $D_n$ for $n \geqslant 3$. What differences are there between the case when $n$ is even and when $n$ is odd?

### EXERCISE 15

Show that the centre of the dihedral group $D_4$ has order 2. Investigate the centre of $D_n$ for other values of $n$.

### EXERCISE 16

Let $G, H$ be isomorphic groups with an isomorphism $f : G \to H$. Show that $Z(H) = f(Z(G))$. Deduce from Exercise 13.9 that $Z(D_3) = \{e\}$.

### EXERCISE 17

Let $a = \operatorname{rot}_\theta$ be a rotation with $0 < \theta < 2\pi$ and $\theta \neq \pi$. Let $G$ be the orthogonal group $O_2$. Show that $\operatorname{cent}_G(a)$ contains all the rotations but contains no reflections. (Thus, $\operatorname{cent}_G(a) = SO_2$.) What are $\operatorname{cent}_G(\operatorname{rot}_0)$ and $\operatorname{cent}_G(\operatorname{rot}_\pi)$?

## EXERCISE 18

Let $a$ be a reflection in the orthogonal group $O_2$. Show that $\text{cent}_{O_2}(a)$ has order four. Is it cyclic or is it isomorphic to Klein's 4-group?

## EXERCISE 19

Show that the centre of the orthogonal group $O_2$ has order 2.

## EXERCISE 20

Let $G$ be a group acting on a set $X$ and let $a, g \in G$. Show that $g * x \in \text{fix}(gag^{-1})$ for all $x \in \text{fix}(a)$ and that $g^{-1} * y \in \text{fix}(a)$ for all $y \in \text{fix}(gag^{-1})$. Hence, show that if $X$ is finite then $|\,\text{fix}(gag^{-1})| = |\,\text{fix}(a)|$.

## EXERCISE 21

Let $G$ be the general linear group $GL_2(\mathbb{R})$ and let $X$ be the set of all $2 \times 2$ real matrices. For $A \in G$ and $M \in X$, let $A * M = AMA^T$.

(i) Show that $*$ is an action of $G$ on $X$.

(ii) Show that the stabilizer of the identity matrix $I_2$ is the orthogonal group $O_2$.

(iii) Let $M = \begin{pmatrix} 1 & 0 \\ 0 & -1 \end{pmatrix}$, let $H = \left\{ \begin{pmatrix} a & b \\ b & a \end{pmatrix} : a, b \in \mathbb{R} \text{ and } a^2 - b^2 = 1 \right\}$ and let $K = \left\{ \begin{pmatrix} a & b \\ -b & -a \end{pmatrix}, \quad a, b \in \mathbb{R} \text{ and } a^2 - b^2 = 1 \right\}$. We have seen in Chapter 5 that $H$ is a subgroup of $GL_2(\mathbb{R})$. Show that $H \subseteq \text{stab}(M)$, that $K \subseteq \text{stab}(M)$ and that $K$ is equal to the left coset $MH$ of $H$ in $\text{stab}(M)$.

(iv) Show that $\text{stab}(M) = H \cup K$ and that $H$ and $K$ are the only left cosets of $H$ in $\text{stab}(M)$.

(v) Let $N = \begin{pmatrix} 1 & 0 \\ 0 & 0 \end{pmatrix}$. Find $\text{stab}(N)$.

# 14 • Towards Classification

We have seen in Chapter 10 that if $p$ is a prime number then all groups of order $p$ are cyclic. This result classifies groups of order $p$. In Chapter 17, we shall classify groups of order $2p$, groups of order $p^2$ and groups of order $\leqslant 12$. The aim of this chapter is to prepare the ground for such classification problems. One way in which a group might be classified is as a direct product of two smaller groups. In the second half of this chapter we shall look at direct products in more detail than before. We shall see how to recognize when a group is isomorphic to a direct product.

## 14.1 An Action of $S_3$ on Three-dimensional Space

On page 112, we saw that every group $G$ of order 15 has an element of order 3 but found that the method used did not show that $G$ must have an element of order 5. To classify groups of order 15, it would be helpful to know whether there must be an element of order 5. We now aim to prove that for any prime number $p$ every finite group $G$ whose order is a multiple of $p$ must contain an element of order $p$. This is called **Cauchy's** Theorem.

In the remainder of this section, we look at an action of the group $S_3$ on three-dimensional space $\mathbb{R}^3$. Cauchy's Theorem will then be proved using a general type of action for which this is a prototype.

For a point $(a, b, c) \in \mathbb{R}^3$ and a permutation $f \in S_3$, the point $f * P$ is determined by the rule that, for $i = 1, 2, 3$, the $i$th coordinate of $P$ becomes the $f(i)$th coordinate of $f * P$. Thus, the coordinates of $P$ are permuted using $f$. For example

$$(1\ 2\ 3) * (a, b, c) = (c, a, b) \text{ and } (1\ 3) * (a, b, c) = (c, b, a).$$

We need to check that this is an action.

**GA1**: Clearly $\text{id} *(a, b, c) = (a, b, c)$.

**GA2**: Let $f, g \in S_3$ and let $P = (a, b, c) \in \mathbb{R}^3$. The $i$th coordinate of $P$ becomes the $g(i)$th coordinate of $g * P$ which then becomes the $fg(i)$th coordinate of $f * (g * P)$. But when $fg$ is applied, the $i$th coordinate of $P$ becomes the $fg(i)$th coordinate of $(fg) * P$. Hence, $f * (g * P) = (fg) * P$.

The orbit of $P = (1, 0, 0)$ consists of $P$, $Q$ and $R$, where $Q = (0, 1, 0)$ and $R = (0, 0, 1)$, while its stabilizer is $H =< (2\ 3) >$. The permutations which send $P$ to $Q$ are $(1\ 2)$ and $(1\ 2\ 3)$; that is, $\text{send}_P(Q) = \{(1\ 2), (1\ 2\ 3)\}$. You can check that, in accordance with the general theory, this is equal to the left coset $(1\ 2)H$. The bijection, $\text{send}_P$, between the orbit of $P$ and the set of left cosets of $H$ is

$$P \mapsto H, \quad Q \mapsto (1\ 2)H, \quad R \mapsto (1\ 3)H.$$

**EXERCISE 1**

Consider the action of $S_3$ on $\mathbb{R}^3$ described above. For each of the following points $P$, find orb($P$), find stab($P$) and, for each $Q$ in the orbit of $P$, find $\text{send}_P(Q)$ .
(i) $(1, 1, 0)$; (ii) $P = (1, 1, 1)$.

## 14.2 Cauchy's Theorem

In order to prove Cauchy's Theorem, we look at a variant of the action in the previous section where the set $\mathbb{R}$ is replaced by the group $G$ and the symmetric group $S_3$ is replaced by $S_p$. Before we consider this in general, we look at it in the case where $p = 3$ and $G$ is the dihedral group $D_3$, where the connection between elements of order 3 and orbits of the action will emerge.

Replacing $\mathbb{R}$ by $D_3$ in Section 14.1, there is an action of the symmetric group $S_3$ on the set $Y$ of all 3-tuples $(g_1, g_2, g_3)$ of elements of $D_3$ whereby, for example,
$(1\ 2\ 3) * (r_1, r_2, e) = (e, r_1, r_2)$.

Now let $A$ be the cyclic subgroup of $S_3$ generated by $(1\ 2\ 3)$ and let $X$ be the set of 3-tuples $(g_1, g_2, g_3)$, where each $g_i \in D_3$ and $g_1g_2g_3 = e$. Examples of members of $X$ are $(r_1, r_2, e)$, $(s_1, e, s_1)$ and, because $r_2$ has order 3, $(r_2, r_2, r_2)$. Notice that in each case we have $g_3 = (g_1g_2)^{-1}$ to ensure that $g_1g_2g_3 = e$. Let $x = (g_1, g_2, g_3) \in X$. Thus, $g_1g_2g_3 = e$. The three elements of $A$ act on $x$ as follows

$$\text{id} * x = x, \quad (1\ 2\ 3) * x = (g_3, g_1, g_2), \quad (1\ 3\ 2) * x = (g_2, g_3, g_1).$$

Then $g_3g_2g_1 = (g_1g_2)^{-1}g_1g_2 = e$ and $g_2g_3g_1 = g_2(g_1g_2)^{-1}g_1 = g_2g_2^{-1}g_1^{-1}g_1 = e$ so

$$f * x \in X \text{ for all } x \in X \text{ and all } f \in A.$$

Thus, we have an action of the subgroup $A$ on the set $X$.

How big is the set $X$? For each of the first two elements $g_1, g_2$ of $(g_1, g_2, g_3)$ there are six choices, since $|D_3| = 6$, but $g_3$ is then forced to be $(g_1g_2)^{-1}$. The order of $X$ is therefore $6 \times 6 = 36$.

The next exercise is intended to help you to understand the key points of the proof of Cauchy's Theorem, which follows the exercise.

**EXERCISE 2**

Consider the action of $A$ on $X$ described above.

(i) Find $g_3 \in D_3$ such that $(r_1, s_1, g_3) \in X$. How many elements of $X$ are in the orbit of $(r_1, s_1, g_3)$?

(ii) Find $g_3 \in D_3$ such that $(r_1, r_1, g_3) \in X$. How many elements of $X$ are in the orbit of $(r_1, r_1, g_3)$?

(iii) Which elements $(g_1, g_2, g_3)$ of $X$ are stabilized by $(1\ 2\ 3)$?

### • Theorem 1 (Cauchy's Theorem)

Let $G$ be a group of order $n$ and let $p$ be a prime divisor of $n$. Then $G$ has an element of order $p$. ●

PROOF

Let $A$ be the cyclic subgroup of $S_p$ generated by $(1\,2 \ldots p)$ and let $X$ be the set of $p$-tuples $(g_1, g_2, \ldots, g_p)$ where each $g_i \in G$ and $g_1 g_2 \ldots g_p = e$. Let $f \in A$ and let $x = (g_1, g_2, \ldots, g_p) \in X$. We check that $f * x \in X$ where $*$ is the action described above. Now $g_p = (g_1 \ldots g_{p-1})^{-1}$ so

$$x = (g_1, \ldots, g_{p-1}, (g_1 g_2 \ldots g_{p-1})^{-1}).$$

Then $f * x$ is obtained from $x$ by cyclically permuting the coordinates and so

$$f * x = (g_i, \ldots, g_{p-1}, (g_1 \ldots g_{p-1})^{-1}, g_1, \ldots, g_{i-1}) \text{ where } f(i) = 1.$$

Now

$$g_i \ldots g_{p-1}(g_1 \ldots g_{p-1})^{-1} g_1 \ldots g_{i-1}$$

$$= (g_i \ldots g_{p-1} g_{p-1}^{-1} \ldots g_i^{-1})(g_{i-1}^{-1} \ldots g_1^{-1} g_1 \ldots g_{i-1}) = e$$

and so $f * x \in X$ as required. Hence, we have an action of $A$ on $X$.

Since $A$ has prime order $p$, we know from Theorem 3 of Chapter 11 that each orbit of this action of $A$ must have either $p$ elements or 1 element. Let $s$ be the number of orbits with $p$ elements and $t$ be the number of orbits with 1 element. The orbits partition $X$ so $X$ has precisely $ps + t$ elements. The size of $X$ can be computed in another way. For each of $g_1, g_2, \ldots, g_{p-1}$ we can choose any of the $n$ elements of $G$ but then we are forced to choose $g_p = (g_1 \ldots g_{p-1})^{-1}$. Thus, $X$ has $n^{p-1}$ elements. As $p$ is a factor of $n$, we have that $n = pm$ for some integer $m$. Hence, $ps + t = |X| = n^{p-1} = (pm)^{p-1}$ and so $t = pq$ where $q = (p^{p-2}m^{p-1} - s)$. Now $t > 0$ because the orbit of $(e, e, \ldots, e)$ has only one element. As $t = pq$ is a positive multiple of $p$ it follows that $t \geqslant p > 1$.

We now know that, apart from $(e, e, \ldots, e)$, there is at least one element, $x = (g_1, g_2, \ldots, g_p)$, whose orbit has only one element. Such an element $x$ is stabilized by $(1\,2 \ldots p)$ so $g_1 = g_2 = \ldots = g_p = g$, say, and $x = (g, g, \ldots, g)$. Because $x \in X$, $g^p = e$ and because $x \neq (e, e, \ldots, e)$, $g \neq e$. The order of $g$ must, by Theorem 3(iii) of Chapter 6, be a divisor of $p$. Since $p$ is prime and $g$ cannot have order 1, the order of $g$ must be $p$. ●

## EXERCISE 3

Let $G$ be a group of order 6.

(i) Show that $G$ has a cyclic subgroup of order 3 and a cyclic subgroup of order 2.

(ii) Suppose that $G$ is abelian. Let $a \in G$ have order 3 and $b \in G$ have order 2. What is the order of $ab$? Show that $G$ is cyclic.

## 14.3 Direct Products

Recall from Chapter 4 that the **direct product** $G \times H$ of two groups $G$ and $H$ is the group of ordered pairs $(g, h)$, where $g \in G$ and $h \in H$, with $(g_1, h_1)(g_2, h_2) = (g_1g_2, h_1h_2)$. In Chapter 9, we saw that Klein's 4-group $V$ is isomorphic to $C_2 \times C_2$, the direct product of two cyclic groups of order 2.

Let $G \times H$ be the direct product of two groups $G$ and $H$. We have two functions, one from each of $G$ and $H$ to $G \times H$

$$i : G \to G \times H; \quad g \mapsto (g, e_H),$$

$$j : H \to G \times H; \quad h \mapsto (e_G, h).$$

The functions $i$ and $j$ are injective homomorphisms. The image of $G$ under $i$ is the set $i(G) = \{(g, e_H) : g \in G\}$ As $i$ is injective, we have an isomorphism from $G$ to $i(G)$

$$g \mapsto (g, e_H).$$

Similarly, $H$ is isomorphic to the subgroup $j(H) = \{(e_G, h) : H\}$. Notice that every element $(g, h)$ of $G \times H$ is a product $i(g)j(h)$ where $i(g) = (g, e_H) \in i(G)$ and $j(h) = (e_G, h) \in j(II)$. We can view $G$ and $H$ as subgroups of $G \times H$ by an abuse of notation whereby we write $(g, e_H)$ as $g$ and $(e_G, h)$ as $h$.

### Multiplying subsets

Let $A$ and $B$ be two non-empty subsets of a group $G$. We define the **product** AB to be the set

$$\{ab : a \in A, b \in B\}.$$

Thus, $AB$ consists of all elements of $G$ which can be written as a product of an element of $A$ and an element of $B$, with the element from $A$ on the left. If $A = B$ we write $A^2$ rather than $AA$.

### • Theorem 2

Let $G$ be a group with subgroups $H$ and $K$ such that $H \cap K = \{e\}$.

(i) If $H$ and $K$ are finite with orders $r, s$ respectively then $HK$ has $rs$ different elements. In particular, if $rs = |G|$ then $G = HK$.

(ii) If $hk = kh$ for all $h \in H$ and all $k \in K$ then $HK$ is a subgroup of $G$ and is isomorphic to the direct product $H \times K$. •

PROOF

(i) Suppose that $h_1k_1 = h_2k_2$. Premultiplying by $h_2^{-1}$ and postmultiplying by $k_1^{-1}$, we obtain $h_2^{-1}h_1 = k_2k_1^{-1}$. But $h_2^{-1}h_1 \in H$ and $k_2k_1^{-1} \in K$ so $h_2^{-1}h_1 = k_2k_1^{-1} \in H \cap K = \{e\}$. Hence, $h_2^{-1}h_1 = e = k_2k_1^{-1}$ and therefore $h_2 = h_1$ and $k_2 = k_1$. Thus, $h_1k_1 = h_2k_2$ only when $h_1 = h_2$ and $k_1 = k_2$. Hence, there are $rs$ distinct elements of the form $hk$, $h \in H, k \in K$.

(ii) We apply the subgroup criterion to $HK$. For SG1, $e = ee$ so $HK$ is non-empty. For SG2, let $h_1k_1, h_2k_2 \in HK$. Then, because $k_1h_2 = h_2k_1$, we have $(h_1k_1)(h_2k_2) = (h_1h_2)(k_1k_2) \in HK$. Thus, SG2 holds. Finally, SG3 holds because $(h_1k_1)^{-1} = (k_1h_1)^{-1} = h_1^{-1}k_1^{-1} \in HK$. Hence, $HK$ is a subgroup of $G$.

By the proof of (i), each element of $HK$ can be written in a unique way in the form $hk, h \in H, k \in K$. The function

$$f : H \times K \to HK, \quad f((h, k)) = hk,$$

is bijective and is a homomorphism because

$$f((h_1, k_1)(h_2, k_2)) = f((h_1h_2, k_1k_2)) = h_1h_2k_1k_2 = h_1k_1h_2k_2 = f((h_1, k_1))f((h_2, k_2)).$$

Thus, $f$ is an isomorphism from $H \times K$ to $HK$. ●

## Example 1

In the symmetric group $S_6$, let $a = (1\ 2\ 3)$ and $b = (4\ 5\ 6)$. These have order 3 and commute, being disjoint cycles of length 3. Let $H =< a >= \{\text{id}, (1\ 2\ 3), (1\ 3\ 2)\}$ and let $K =< b >= \{\text{id}, (4\ 5\ 6), (4\ 6\ 5)\}$. Then $H \cap K = \{\text{id}\}$ and each element of $H$ commutes with each element of $K$. Hence the nine elements

$$\begin{array}{ccc} \text{id} & a = (1\ 2\ 3) & a^2 = (1\ 3\ 2) \\ b = (4\ 5\ 6) & ab = (1\ 2\ 3)(4\ 5\ 6) & a^2b = (1\ 3\ 2)(4\ 5\ 6) \\ b^2 = (4\ 6\ 5) & ab^2 = (1\ 2\ 3)(4\ 6\ 5) & a^2b^2 = (1\ 3\ 2)(4\ 6\ 5) \end{array}$$

form a subgroup of $S_6$ isomorphic to $C_3 \times C_3$.

## Example 2

Let $G$ be an abelian group of order 15. By Cauchy's Theorem, $G$ has an element $a$ of order 3 and an element $b$ of order 5. If $H =< a >$ and $K =< b >$ then by Theorem 9(i) of Chapter 10, $H \cap K = \{e\}$. It follows from Theorem 2 that $G = HK \cong H \times K$. By Theorem 9 of Chapter 6, $G$ must be cyclic.

## Example 3

Consider the two groups $D_6$ and $D_3 \times C_2$. Both have order 12 and neither is abelian. If you count the number of elements of each order $1, 2, 3, 4, 6, 12$ in either, you get the same answers $1, 7, 2, 0, 2, 0$. This suggests that maybe these two groups are isomorphic. To see that they are, consider the subgroups $H = D_3$ and $K =< \text{rot}_\pi >$ of $D_6$. Thus, $|H| = 6$ and $|K| = 2$. Note that $\text{rot}_\pi \notin H$ because the rotations in $H$ are $e$, $\text{rot}_{\frac{2\pi}{3}}$ and $\text{rot}_{\frac{4\pi}{3}}$. Therefore, $H \cap K = \{e\}$. Furthermore, $\text{rot}_\pi$ commutes with all rotations and reflections and hence with every element of $H$. By Theorem 2, every element of $D_6$ must be of the form $hk, h \in D_3, k \in K$ and $D_6 \cong H \times K$. As $K$ is cyclic of order 2, we can write $H \times K$ as $D_3 \times C_2$.

The language of isomorphism is formal and does not explain the geometry. The underlying geometric reason for the isomorphism between $D_6$ and $D_3 \times C_2$ is that the

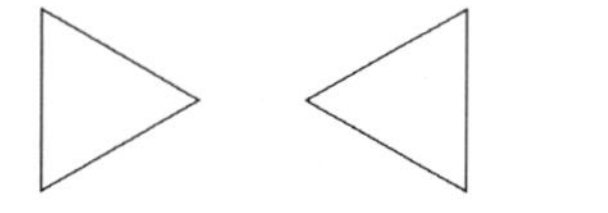
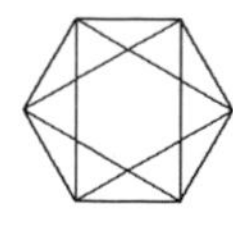

**Fig 14.1** Two triangles make a hexagon.

vertices of a triangle and its image under rotation through $\pi$ are the vertices of the hexagon; see Fig 14.1.

There are various situations where two direct products are easily seen to be isomorphic.

- If $G_1 \cong G_2$ and $H_1 \cong H_2$, with isomorphisms $f_1 : G_1 \to G_2$ and $f_2 : H_1 \to H_2$, then $G_1 \times H_1 \cong G_2 \times H_2$ with an isomorphism sending $(g, h)$ to $(f_1(g), f_2(h))$.
- There is an isomorphism $f : G \times H \to H \times G$ with $f : (g, h) \mapsto (h, g)$. Thus, $G \times H \cong H \times G$.
- There is an isomorphism $j : (G \times H) \times K \to G \times (H \times K)$ with $j : ((g, h), k) \mapsto (g, (h, k))$. Thus, $(G \times H) \times K \cong G \times (H \times K)$.

The last of these tells us that we can omit brackets when writing the direct product $G \times H \times K$ of three groups. For example $C_2 \times C_2 \times C_2$ is a group of order 8 in which every element is its own inverse.

## Example 4

The elements of $Z_{24}$ which have inverses for multiplication modulo 24 are $\overline{1}, \overline{5}, \overline{7}, \overline{11}, \overline{13}, \overline{17}, \overline{19}$ and $\overline{23}$. It can be checked that these are all their own inverses. Thus, the group $U(Z_{24}) = \{\overline{1}, \overline{5}, \overline{7}, \overline{11}, \overline{13}, \overline{17}, \overline{19}, \overline{23}\}$ is an abelian group of order 8 with seven elements of order 2. Let $H =< \overline{5} >$ and $K =< \overline{7} >$. Then the conditions of Theorem 2(ii) are satisfied and we get a subgroup $P = HK = \{\overline{1}, \overline{5}, \overline{7}, \overline{11}\}$ isomorphic to $C_2 \times C_2$, consisting of all elements $hk$ with $h \in H$ and $k \in K$. Now consider the subgroups $P$ and $L =< \overline{13} >$. These also satisfy the conditions of Theorem 2(ii) and so we get a subgroup $PL$ of order 8, necessarily the whole group, isomorphic to $P \times L$ and hence to $C_2 \times C_2 \times C_2$. Thus, $U(\mathbb{Z}_{24}) \cong C_2 \times C_2 \times C_2$.

## EXERCISE 4

Let $G$ be a group with two distinct commuting elements $a$ and $b$ of order 2. Let $c = ab = ba$ and let $V$ be the subset $\{e, a, b, c\}$ of $G$.

(i) Show that $V$ is a subgroup of $G$ isomorphic to $C_2 \times C_2$.

(ii) Show that if $G$ is finite then $|G|$ is a multiple of 4.

## EXERCISE 5

By considering elements of order 8, show that $D_8$ is not isomorphic to $D_4 \times C_2$.

### EXERCISE 6

The group $U(\mathbb{Z}_{15})$ has order 8, its elements being $\overline{1}, \overline{2}, \overline{4}, \overline{7}, \overline{8}, \overline{11}, \overline{13}$ and $\overline{14}$. Show that $U(\mathbb{Z}_{15}) \cong C_4 \times C_2$.

## 14.4 Further Exercises on Chapter 14

### EXERCISE 7

Let $G$ be an abelian group of order 21. Show that $G$ is cyclic.

### EXERCISE 8

What can you say about an abelian group of order $pq$ where $p$ and $q$ are distinct primes?

### EXERCISE 9

Show that a group of order 15 has at most ten elements of order 3.

### EXERCISE 10

Find a subgroup of $S_4$ isomorphic to $C_2 \times C_2$ and a subgroup of $S_6$ isomorphic to $C_4 \times C_2$.

### EXERCISE 11

Investigate non-abelian groups of order 6. Are there any which are not isomorphic to $S_3$? The answer will be revealed in Chapter 17.

### EXERCISE 12

(i) Let $G$ be any group. Show that the direct product $G \times C_2$ has a central element of order 2.

(ii) Investigate for which values of $n$, the dihedral group $D_n$ is isomorphic to a group of the form $G \times C_2$.

### EXERCISE 13

Investigate $U(\mathbb{Z}_n)$, for different values of $n$, from the point of view of direct products of cyclic groups.

## EXERCISE 14

Let $G$ and $H$ be groups. Consider the functions

$$p_1 : G \times H \to G; \quad (g, h) \mapsto g,$$

$$p_2 : G \times H \to H; \quad (g, h) \mapsto h.$$

(In the case where $G = H = \mathbb{R}$, the function $p_1 : (x, y) \mapsto x$ is just the projection onto the $x$-axis and the name **projection** is given to the functions $p_1$, $p_2$ in general.)

(i) Show that $p_1$ and $p_2$ are homomorphisms.

(ii) Discuss whether these homomorphisms are injective/surjective.

## EXERCISE 15

Let $G$ be any group and let $f : G \to G \times G$ be the function

$$f : g \mapsto (g, g).$$

Show that $f$ is a homomorphism. What is its image?

## EXERCISE 16

Let $G$ be a group. Show that the function $f : G \times G \to G$ given by $f : (g_1, g_2) \mapsto g_1 g_2$ is a homomorphism if and only if $G$ is abelian.

## EXERCISE 17

Show that the centre of a direct product $G \times H$ is $Z(G) \times Z(H)$.

# 15 • Kernels and Normal Subgroups

The stabilizer of any element $x$ of a set $X$ under a group action $*$ consists of those group elements which send $x$ to itself. Related to this idea is that of the kernel of a group action $*$. This subgroup consists of those group elements that send every element of $X$ to itself. We shall define kernels first for homomorphisms and then use the homomorphism arising from an action to define the kernel of the action. The subgroups that can occur as kernels behave well with regard to conjugates and are called normal subgroups.

## 15.1 Kernels of Homomorphisms

Consider the homomorphism $f : D_4 \to S_2$ arising from the action of the dihedral group $D_4$ on the set $\{1, 2\}$ of numbered diagonals of the square. Thus, for $g \in D_4$, $f(g) = f_g$, the permutation of the diagonals corresponding to $g$. There are only two possibilities for $f(g)$, namely id and $(1\ 2)$. Each element of $D_4$ either stabilizes both diagonals or swops them over. This gives a partition of $D_4$ into two subsets, each of four elements

$$K = \{e, r_2, s_2, s_4\}; \quad L = \{r_1, r_3, s_1, s_3\}.$$

The four elements $g$ of $K$ are distinguished by the property that $f(g) = \text{id} = e_{S_2}$.

Let $f : G \to H$ be a homomorphism. The **kernel** of $f$, written $\ker f$, is the set

$$\ker f = \{g \in G : f(g) = e_H\}.$$

Thus, $\ker f$ consists of all those elements of $G$ which are sent to the neutral element of $H$ by $f$. With $f : D_4 \to S_2$ as above, $\ker f = K$.

If $H$ is in additive notation then $\ker f = \{g \in G : f(g) = 0\}$.

### Example 1

Consider the homomorphism $f : GL_2(\mathbb{R}) \to \mathbb{R}\backslash\{0\}$ such that $f(A) = \det A$ for all $A \in GL_2(\mathbb{R})$. Then $\ker f$ consists of those elements $A$ of $GL_2(\mathbb{R})$ with $\det A = 1$. Thus, $\ker f$ is the special linear group $SL_2(\mathbb{R})$.

### Example 2

The alternating group $A_n = \{\alpha \in S_n : \operatorname{sgn} \alpha = +1\}$ is the kernel of the homomorphism $f : S_n \to \mathbb{R}\backslash\{0\}$ given by $f : \alpha \mapsto \operatorname{sgn} \alpha$.

In the above examples, the kernel of $f : G \to H$ is a subgroup of $G$. This is no accident.

### • Theorem 1

Let $f : G \to H$ be a homomorphism. Then $\ker f$ is a subgroup of $G$. •

PROOF

We use the subgroup criterion.

**SG1:** By Theorem 2(i) of Chapter 9, $f(e_G) = e_H$ so $e_G \in \ker f$, whence $\ker f \neq \emptyset$.

**SG2:** Let $a, b \in \ker f$. Thus, $f(a) = e = f(b)$. Then $f(ab) = f(a)f(b) = ee = e$ so $ab \in \ker f$.

**SG3:** If $a \in \ker f$ then, by Theorem 2(ii) of Chapter 9, $f(a^{-1}) = (f(a))^{-1} = e^{-1} = e$ so $a^{-1} \in \ker f$. ●

It is possible for the kernel of a homomorphism to be the trivial subgroup $\{e\}$. The significance of this is given by the next theorem.

### ● Theorem 2

Let $f : G \to H$ be a homomorphism. Then $f$ is injective if and only if $\ker f = \{e_G\}$. ●

PROOF

First suppose that $f$ is injective and let $a \in \ker f$. Then $f(a) = e_H = f(e_G)$ so, by injectivity, $a = e_G$. Thus, $\ker f = \{e_G\}$.

Conversely, suppose that $\ker f = \{e_G\}$ and let $a, b \in G$ be such that $f(a) = f(b)$. Then

$$f(ab^{-1}) = f(a)f(b^{-1}) = f(a)(f(b))^{-1} = f(b)(f(b))^{-1} = e_H$$

so $ab^{-1} \in \ker f = \{e_G\}$. Thus, $ab^{-1} = e_G$ and so $a = b$. Therefore, $f$ is injective. ●

## EXERCISE 1

Find the kernels of the following homomorphisms and decide which of the homomorphisms are injective.

(i) $f : \mathbb{Z} \to \mathbb{Z}_n;\ f : a \mapsto \overline{a}$.

(ii) $f : \mathbb{R} \to O_2;\ f : \theta \mapsto \text{rot}_\theta$.

(iii) $f : \mathbb{R} \to GL_2(\mathbb{R});\ a \mapsto \begin{pmatrix} 1 & a \\ 0 & 1 \end{pmatrix}$.

(iv) $f : O_2 \to O_2;\ f : \begin{cases} \text{rot}_\theta \mapsto \text{rot}_{2\theta}, \\ \text{ref}_\theta \mapsto \text{ref}_{2\theta} \end{cases}$.

## 15.2 Kernels of Actions

Arising from any action $*$ of a group $G$ on a set $X$, is the homomorphism $f : G \to S_X$ where $f : g \mapsto f_g$, the permutation of $X$ corresponding to $g$. The kernel of this homomorphism is also called the **kernel** of the action $*$ and is written $\ker *$. Remember that $f(g) = f_g$ is the permutation which sends each $x$ to $g * x$. Therefore, $f(g) = \text{id}_X$ if and only if $g * x = x$ for all $x \in X$ and so

$$\ker * = \{g \in G : g * x = x \text{ for all } x \in X\}$$

consists of those elements of $G$ which stabilize *every* element of $X$.

By Theorem 1, ker $*$ is always a subgroup of $G$. For the action of $D_4$ on the diagonals of the square, ker $* = \{e, r_2, s_2, s_4\}$. Apart from $e$, every symmetry of the square moves at least one vertex so, for the action of $D_4$ on the *vertices* of the square, the kernel is $\{e\}$, the trivial subgroup.

Let $*$ be an action of a group $G$ on a set $X$. Then we say that $*$ is **faithful**, or that $G$ acts **faithfully** on $X$, if ker $* = \{e\}$. Thus, $*$ is faithful if and only if the *only* element of $G$ which sends every element of $X$ to itself is $e$. The action of $D_4$ on the diagonals of the square is not faithful but the action of $D_4$ on the vertices is faithful.

## EXERCISE 2

For each of the following actions of $D_4$ (see Fig 7.1) find the kernel of the action and state whether the action is faithful.

(i) The action of $D_4$ on the sides of the square.

(ii) The action of $D_4$ on the two axes.

# 15.3 Conjugates of a Subgroup

Let $H$ be a subgroup of a group $G$ and let $a \in G$. The **conjugate** of $H$ by $a$, written $aHa^{-1}$, is the set

$$aHa^{-1} = \{aha^{-1} : h \in H\}$$

of all conjugates of elements of $H$ by $a$. This is the image $f(H)$ of $H$, under the conjugation homomorphism $f : G \to G$ where $f : g \mapsto aga^{-1}$. Hence, $aHa^{-1}$ is a subgroup of $G$.

### Example 3

Let $G = D_4$ and let $a = r_1$. We compute the conjugates $aHa^{-1}$ of two subgroups $H$ of $G$. First, let $H$ be the subgroup $\{e, r_2, s_2, s_4\}$, a copy of Klein's 4-group. Then, using the Cayley table for $D_4$, we have

$$aea^{-1} = e,\ ar_2a^{-1} = r_2,\ as_2a^{-1} = s_4,\ as_4a^{-1} = s_2 \text{ so } aHa^{-1} = \{e, r_2, s_4, s_2\} = H.$$

However, if $H$ is the cyclic subgroup $< s_2 >= \{e, s_2\}$ then $aHa^{-1} = \{e, s_4\} =< s_4 >\neq H$.

This example shows that a conjugate of a given subgroup $H$ might or might not equal $H$. It will be important to know when $H = aHa^{-1}$ for all $a \in G$. Usually when we have to show two sets to be equal, we have to show that each is inside the other. However, the next theorem shows that here we need only check one inclusion.

### Theorem 3

Let $H$ be a subgroup of a group $G$. If $aHa^{-1} \subseteq H$ for all $a \in G$ then $aHa^{-1} = H$ for all $a \in G$.

PROOF

Suppose that $aHa^{-1} \subseteq H$ for all $a \in G$. Let $h \in H$ and let $a \in G$. Then $a^{-1} \in G$ so $a^{-1}ha = a^{-1}h(a^{-1})^{-1} \in H$, that is $a^{-1}ha = h_1$ for some $h_1 \in H$. But then $h = ah_1a^{-1} \in aHa^{-1}$. Thus, $H \subseteq aHa^{-1}$ and so, as the reverse inclusion also holds, $H = aHa^{-1}$. ●

We now show that the conjugates of a subgroup $H$ of $G$ occur as stabilizers of a group action of $G$. Let $H$ be a subgroup of a group $G$ and let $X = \{aH : a \in G\}$ be the set of all left cosets of $H$ in $G$. For $g \in G$ and $aH \in X$, let

$$g * aH = gaH \in X.$$

To check that this is a group action, let $aH \in X$. Then $e * aH = eaH = aH$ and, for all $g, h \in G$, $g * (h * aH) = g * haH = g(ha)H = (gh)aH = (gh) * aH$. Thus, $*$ is an action of $G$ on $X$. We aim to find the stabilizers for this action.

## Example 4

The centre of $D_4$ is $H = \langle r_2 \rangle = \{e, r_2\}$. Its four different left cosets are

$$L_1 = H,\ L_2 = r_1H = \{r_1, r_3\},\ L_3 = s_1H = \{s_1, s_3\} \text{ and } L_4 = s_2H = \{s_2, s_4\}.$$

Let $f : D_4 \to S_4$ be the homomorphism arising from the action of $D_4$ on the left cosets by the rule $g * aH = gaH$. Then $f(e) = \text{id}$ and $f(r_1) = (1\ 2)(3\ 4)$ because $r_1 * L_1 = L_2,\ r_1 * L_2 = L_1,\ r_1 * L_3 = L_4$ and $r_1 * L_4 = L_3$. As $f$ is a homomorphism, $f(r_2) = f(r_1^2) = (f(r_1))^2 = \text{id}$ and $f(r_3) = f(r_1)f(r_2) = (1\ 2)(3\ 4)$. Similar calculations show that $f(s_1) = (1\ 3)(2\ 4) = f(s_3)$ and $f(s_2) = (1\ 4)(2\ 3) = f(s_4)$. Thus, the kernel is $H$. Notice that, for each of the left cosets, $f$ has the same effect on the two elements of the coset. The image $f(D_4)$ is $\{\text{id}, (1\ 2)(3\ 4), (1\ 3)(2\ 4), (1\ 4)(2\ 3)\}$, which is a copy of Klein's 4-group.

## ● Theorem 4

Let $G$ be a group acting on the set $X = \{aH : a \in G\}$ of left cosets of a subgroup $H$ of $G$.

(i) The stabilizer of each left coset $aH$ is the conjugate $aHa^{-1}$.

(ii) If $H = aHa^{-1}$ for all $a \in G$ then $H$ is the kernel of the action. ●

PROOF

(i) First let $aha^{-1} \in aHa^{-1}$. Then

$$aha^{-1} * aH = aha^{-1}aH = ahH = aH.$$

Thus, $aha^{-1} \in \text{stab}(aH)$ and so $aHa^{-1} \subseteq \text{stab}(aH)$.

Now let $g \in \text{stab}(aH)$. Then $gaH = g * aH = aH$ so, by Theorem 4(i) of Chapter 10, $a^{-1}ga \in H$. Thus, $a^{-1}ga = h$ for some $h \in H$. But then $g = aha^{-1} \in aHa^{-1}$. Therefore, $\text{stab}(aH) \subseteq aHa^{-1}$ and, as we already have the reverse inclusion, $\text{stab}(aH) = aHa^{-1}$.

(ii) Suppose that $H = aHa^{-1}$ for all $a \in G$. By (i), $H \subseteq \text{stab}(aH)$ for all $aH \in X$ and so $H \subseteq \ker *$. Also by (i), $\text{stab}(eH) = H$ so $\ker * \subseteq \text{stab}(eH) = H$. As we have both inclusions, $H = \ker *$. ●

Subgroups which are equal to all their own conjugates are the subject of the next section.

**EXERCISE 3**

Compute $aHa^{-1}$ for each of the following

(i) $G = D_4, H =< s_1 >, a = r_3$;

(ii) $G = D_4, H =< s_1 >, a = s_3$;

(iii) $G = D_4, H = \{e, r_2, s_2, s_4\}, a = r_3$.

## 15.4 Normal Subgroups

Let $N$ be a subgroup $N$ of a group $G$. We say that $N$ is a **normal** subgroup of $G$, or that $N$ is normal in $G$, if $gNg^{-1} = N$ for all $g \in G$.

By Theorem 3, $gNg^{-1} = N$ for all $g \in G$ if and only if $gNg^{-1} \subseteq N$ for all $g \in G$ so $N$ is normal in $G$ if and only if $gng^{-1} \in N$ for all $g \in G$ and all $n \in N$.

Another formulation of normality is in terms of left and right cosets.

### ● Theorem 5

The subgroup $N$ is normal in $G$ if and only if, for all $g \in G$, the left and right cosets $gN$ and $Ng$ are equal. ●

PROOF

This follows immediately from the observation that $gN = Ng$ if and only if $gNg^{-1} = N$. ●

The condition $gN = Ng$ for all $g \in G$ is sometimes given as the definition of normality. For normal subgroups, left and right cosets are the same so we can simply write coset rather than left or right coset.

### Example 5

Let $G$ be the orthogonal group and let $N$ be its subgroup $SO_2$, which consists of all the rotations. Let $g \in G$ and let $n \in N$. Then $n = \text{rot}_\theta$ for some $\theta$. We have seen in Section 13.1 that if $g$ is a reflection then $gng^{-1} = \text{rot}_{-\theta}$, whereas if $g$ is a rotation then $gng^{-1} = \text{rot}_\theta$. In both cases $gng^{-1} \in N$. Thus, $SO_2$ is a normal subgroup of $O_2$.

### Example 6

We observe here that if $G$ is any group $G$ then the trivial subgroup $\{e\}$ and the whole group $G$ are normal subgroups of $G$. For the trivial subgroup, normality holds because $geg^{-1} = e$ for all $g \in G$. For the whole group, it is certainly true that $gng^{-1} \in G$ for all $g, n \in G$.

It is a consequence of Theorem 4 that if $N$ is a normal subgroup of $G$ then $N$ is the kernel of the action of $G$ on the set $X$ of cosets of $N$ in $G$. Hence, it is also the kernel of a homomorphism, namely the homomorphism from $G$ to $S_X$ arising from this action. We now show that kernels of homomorphisms are always normal.

## Theorem 6

Let $G$ and $H$ be groups. The kernel of any homomorphism $f : G \to H$ is normal in $G$. Hence, the kernel of any group action of $G$ is normal in $G$.

PROOF

Let $N = \ker f$ and let $g \in G, n \in N$. Then $f(n) = e$ and

$$f(gng^{-1}) = f(g)f(n)f(g^{-1}) = f(g)e(f(g))^{-1} = e,$$

using Theorem 2(ii) of Chapter 9 which says that $f(g^{-1}) = (f(g))^{-1}$. Thus, $gng^{-1} \in \ker f = N$ and so $N$ is normal. By definition, the kernel of an action of $G$ on $X$ is the kernel of the corresponding homomorphism from $G$ to $S_X$ and so is normal.

## Example 7

(i) By Example 2, the alternating group $A_n$ is normal in $S_n$, being the kernel of the homomorphism $f : \alpha \mapsto \operatorname{sgn} \alpha$.

(ii) The special linear group $SL_n(\mathbb{R})$, being the kernel of the homomorphism $\det : GL_n(\mathbb{R}) \to \mathbb{R}\backslash\{0\}$, is normal in the general linear group $GL_n(\mathbb{R})$.

## Theorem 7

Let $G$ be an abelian group and let $N$ be any subgroup of $G$. Then $N$ is normal in $G$.

PROOF

Let $g \in G$ and $n \in N$. Then $gng^{-1} = n$ so $N$ is normal.

## EXERCISE 4

Let $G$ be the orthogonal group $O_2$.

(i) Let $N$ be a normal subgroup of $G$ containing a reflection. Show that $N$ contains all the reflections and hence that $N = G$.

(ii) Let $H = \langle \mathrm{rot}_\theta \rangle$ be the cyclic subgroup of $G$ generated by a rotation. Show that $H$ is normal in $G$.

## EXERCISE 5

Show that, in any group $G$, the centre, $Z(G)$, is a normal subgroup of $G$.

## Subgroups of order or index 2

Let $G$ be a group and let $H$ be a subgroup of $G$ of order 2. Thus, $H =< a >= \{e, a\}$ for some $a \in G$ of order 2. If $a$ is central in $G$ then, for all $g \in G$, $gHg^{-1} = \{e, a\} = H$ so $H$ is normal in $G$. If $a$ is not central then there exists $g \in G$ with $ga \neq ag$ and hence with $gag^{-1} \neq a$. In this case, $gHg^{-1} = \{e, gag^{-1}\} \neq H$ so $H$ is not normal. Thus, when $a$ has order 2

$< a >$ is normal if and only if $a \in Z(G)$.

For example, $D_3$ has three elements of order 2 but none of these are central. So none of the three subgroups of $D_3$ of order 2 are normal. On the other hand, $r_2 = \text{rot}_\pi$ is a central element of order two in $D_4$ so $D_4$ has a normal subgroup of order two.

### *Subgroups of index 2*

Let $G$ be a group and let $H$ be a subgroup of $G$. We say that $H$ has **index** 2 in $G$ if there are only two different left cosets of $H$ in $G$. There are then only two different right cosets of $H$ in $G$ because there is a bijection pairing left cosets with right cosets, see Section 10.6. When $G$ is finite, $H$ has index 2 when $|G| = 2|H|$. For example, $|S_n| = 2|A_n|$ so $A_n$ has index 2 in $S_n$. It is also possible for an infinite group to have a subgroup of index 2.

### Example 8

Let $G$ be the orthogonal group $O_2$ and let $H$ be the special orthogonal group $SO_2$. Since every reflection has the form $\text{ref}_\beta = \text{ref}_0 \, \text{rot}_{-\beta}$, the only left cosets of $H$ in $G$ are $H$ (the rotations) and $\text{ref}_0 H$ (the reflections). Thus, $SO_2$ has index 2 in $O_2$.

We have seen that $A_n$ and $SO_2$ are normal in $S_n$ and $O_2$ respectively. These are special cases of the next theorem.

### Theorem 8

If $N$ is a subgroup of $G$ of index 2 then $N$ is normal in $G$.

PROOF

The two left cosets partition $G$ so, as one is $N$, the other is the set $G \backslash N$ of all elements of $G$ which are not in $N$. Similarly, the two right cosets must be $N$ and $G \backslash N$. If $a \in N$ then $aN = N = Na$ and if $a \notin N$ then $aN = G \backslash N = Na$. By Theorem 5, $N$ is normal.

### Example 9

The subgroup $H =< r_1 >$ has order 4 in $D_4$ which has order 8. Hence it has index 2 and is normal.

## EXERCISE 6

Let $n \geqslant 3$. Show that the symmetric group $S_n$ has no normal subgroup of order 2 and that $S_n$ has a normal subgroup of index 2.

# 15.5 Normal Subgroups and Conjugacy Classes

Let $N$ be a normal subgroup of a group $G$. If $n \in N$ then all conjugates of $n$ in $G$ must be in $N$ so $\text{conj}_G(n) \subseteq N$. Hence, $N$ is a union of conjugacy classes of $G$. As different conjugacy classes of $G$ do not overlap, this gives a partition of $N$ into conjugacy classes of $G$. If $G$ is finite then $N$ must have the form $N = C_1 \cup C_2 \cup \ldots \cup C_k$, where $C_1, C_2, \ldots, C_k$ are the conjugacy classes of $G$ contained in $N$. It then follows that $|N| = |C_1| + |C_2| + \ldots + |C_k|$ and, because $e \in N$ and $\text{conj}_G\, e = \{e\}$, at least one of the summands is 1. This observation can be quite powerful in analysing the normal subgroups of a given group.

## Example 10

Consider the symmetric group $S_4$ for which the conjugacy classes correspond to cycle types and are shown in Table 15.1.

| $C_i$ | Cycle type | Typical element | $\lvert C_i \rvert$ |
|---|---|---|---|
| $C_1$ | (1, 1, 1, 1) | id | 1 |
| $C_2$ | (2, 1, 1) | (1 2) | 6 |
| $C_3$ | (2, 2) | (1 2)(3 4) | 3 |
| $C_4$ | (3, 1) | (1 2 3) | 8 |
| $C_5$ | (4) | (1 2 3 4) | 6 |

**Table 15.1** Conjugacy classes of $S_4$.

By Lagrange's Theorem, the possible orders of subgroups of $S_4$ are $1, 2, 3, 4, 6, 8, 12$ and 24. Of these, the only ones which can be written as a sum of numbers from the list $1, 6, 3, 8, 6$ with 1 included are 1, 4=1+3, 12=1+3+8, and 24=1+6+3+8+6. So the only possible normal subgroups of $S_4$ are the trivial subgroup {id}, the whole group $S_4$ and the unions $C_1 \cup C_3$ and $C_1 \cup C_3 \cup C_4$. Of course, there is no guarantee that if we take a union of conjugacy classes we have a subgroup. However, here both $C_1 \cup C_3$ and $C_1 \cup C_3 \cup C_4$ are subgroups that we have already met. The latter, $C_1 \cup C_3 \cup C_4$, consists of all the even permutations in $S_4$ and so it is just the alternating group $A_4$. The former is $V = \{\text{id}, (1\,2)(3\,4), (1\,3)(2\,4), (1\,4)(2\,3)\}$, a copy of Klein's 4-group. Notice that $\{\text{id}, (1\,3), (1\,3)(2\,4), (2\,4)\}$, another copy of Klein's 4-group inside $S_4$, is not normal in $S_4$ because it contains only two of the six conjugates of (1 3).

Thus, the normal subgroups of $S_4$ are $S_4$, $A_4$, $V$ and {id}. As any subgroup of $S_4$ of order 12 has index 2 it must be normal. Hence, $A_4$ is the only subgroup of $S_4$ of order 12.

### Conjugacy classes in the alternating groups

We have just seen that $A_4$ is the union of three conjugacy classes in $S_4$ containing 1, 3 and 8 elements respectively. You might think that the class equation for $A_4$ is 12=1+3+8. However, this cannot be correct because the number of elements in a conjugacy class of $A_4$ must, by Theorem 4 of Chapter 13, be a factor of $|A_4| = 12$. Thus, $A_4$ can have no conjugacy class with eight elements. We need to distinguish between conjugacy classes of $A_4$ and conjugacy classes of $S_4$ contained in $A_4$. We shall see that the conjugacy class

$C_4$ of $S_4$ splits into two conjugacy classes, each with four elements, in $A_4$. Thus, the class equation for $A_4$ is 12=1+3+4+4. The conjugacy classes for $A_n$ in general are determined by the next theorem.

## Theorem 9

For $\alpha \in A_n$, let $C_A = \text{conj}_{A_n}(\alpha)$ and $C_S = \text{conj}_{S_n}(\alpha)$ be the conjugacy classes of $\alpha$ in $A_n$ and $S_n$, respectively.

(i) Suppose that $\text{cent}_{S_n}(\alpha) \nsubseteq A_n$, that is, there is at least one odd permutation in $S_n$ commuting with $\alpha$. Then $C_A = C_S$.

(ii) Suppose that $\text{cent}_{S_n}(\alpha) \subseteq A_n$, that is, all permutations in $S_n$ commuting with $\alpha$ are even. Then $|C_A| = \frac{1}{2}|C_S|$.

PROOF

Because $A_n \subseteq S_n$, any conjugate $\beta\alpha\beta^{-1}$ of $\alpha$ in $A_n$ is a conjugate of $\alpha$ in $S_n$. Therefore, $C_A \subseteq C_S$ for all $\alpha \in A_n$.

(i) Let $\tau$ be an odd permutation in $S_n$ with $\alpha\tau = \tau\alpha$. Thus, $\tau\alpha\tau^{-1} = \alpha$. Let $\gamma = \beta\alpha\beta^{-1}$ be a conjugate of $\alpha$ in $S_n$. If $\beta$ is even, that is if $\beta \in A_n$, then $\gamma$ is a conjugate of $\alpha$ in $A_n$. On the other hand, if $\beta$ is odd then $\beta\tau$ is even and

$$(\beta\tau)\alpha(\beta\tau)^{-1} = \beta(\tau\alpha\tau^{-1})\beta^{-1} = \beta\alpha\beta^{-1} = \gamma.$$

Thus, in all cases, $\gamma$ is conjugate to $\alpha$ in $A_n$, that is $\gamma \in C_A$. Hence $C_S \subseteq C_A$ and, as the reverse inclusion also holds, $C_S = C_A$.

(ii) Suppose that $\text{cent}_{S_n}(\alpha) \subseteq A_n$. Then the centralizers, $\text{cent}_{S_n}(\alpha)$ and $\text{cent}_{A_n}(\alpha)$, of $\alpha$ in $S_n$ and $A_n$ are equal. Let $H = \text{cent}_{S_n}(\alpha) = \text{cent}_{A_n}(\alpha)$. By Theorem 4 of Chapter 13 applied in $S_n$ and in $A_n$

$$|C_S| \times |H| = n! \text{ and } |C_A| \times |H| = \frac{n!}{2}.$$

Hence $|C_A| = \frac{1}{2}|C_S|$.

## Example 11

Consider $A_4$ which, as we have seen, is the union of three conjugacy classes $C_1, C_3, C_4$ of $S_4$. Clearly $C_1 = \{\text{id}\} = \text{conj}_{A_4}(\text{id})$ remains a conjugacy class in $A_4$. Consider the element $\alpha = (1\ 2)(3\ 4)$ of $C_3$. Then $\alpha$ commutes with the odd permutation $(1\ 2)$ so, by Theorem 9(i), it has the same conjugacy class in $A_4$ as in $S_4$. Thus, $C_3$ is a conjugacy class in $A_4$.

Now consider $C_4$ and one of its elements, $\alpha = (1\ 2\ 3)$. Taking $n = 4$ in Example 3 of Chapter 13, we see that $\text{cent}_{S_4}(\alpha) =< \alpha >\subseteq A_4$ so, by Theorem 9(ii), only four of the eight conjugates of $\alpha$ in $S_4$ are conjugates of $\alpha$ in $A_4$. To find $\text{conj}_{A_4}(\alpha)$, we compute conjugates $\beta\alpha\beta^{-1}$ for sufficient choices of $\beta \in A_4$ to give four different conjugates. This gives

$$\text{conj}_{A_4}((1\ 2\ 3)) = \{(1\ 2\ 3), (2\ 4\ 3), (1\ 3\ 4), (1\ 4\ 2)\}.$$

Similarly any other 3-cycle has four conjugates in $A_4$, for example

$$\text{conj}_{A_4}(1\ 3\ 2) = \{(1\ 3\ 2), (3\ 4\ 2), (1\ 2\ 4), (1\ 4\ 3)\}.$$

These two classes between them give all of $C_4$ which thus splits into two conjugacy classes in $A_4$. So the class equation for $A_4$ is 12=1+3+4+4.

The converse to Lagrange's Theorem, if true, would say that if $G$ is a finite group and $n$ is a positive integer, which is a factor of $|G|$, then $G$ has a subgroup of order $n$. The next exercise shows that this is not true in general.

## EXERCISE 7

Use the class equation for $A_4$ to show that $A_4$ has exactly three normal subgroups and that these have orders 1,4,12. Hence, show that $A_4$ has no subgroup of order 6. (Hint: subgroups of index 2 are normal.)

# 15.6 Simple Groups

One might expect that as we increase $n$, $A_n$ will have more normal subgroups. In fact this is not so. We shall see below that $A_5$ has only two normal subgroups, the whole group and the trivial subgroup. The same is true for $n > 5$ although we shall not prove that here.

A group $G$ is said to be **simple** if $G \neq \{e\}$ and the only normal subgroups of $G$ are $G$ and $\{e\}$.

Simple groups are important in group theory. In a sense, finite simple groups are the building blocks for all finite groups. It is only recently that all finite simple groups have been classified. The list includes the alternating groups $A_n$ for $n \geqslant 5$, together with other infinite families of groups and 26 so-called sporadic groups which do not fit into any of these families. The proof that the list is complete runs to hundreds of pages and is well beyond the scope of this book.

To prove that $A_5$ is simple, we use the class equation for $A_5$. For this, we first need the conjugacy classes for $S_5$ contained in $A_5$. These are given in Table 15.2. The union of

| $C_i$ | Cycle type | Typical element | $\lvert C_i \rvert$ |
|---|---|---|---|
| $C_1$ | (1, 1, 1, 1, 1) | id | 1 |
| $C_2$ | (2, 2, 1) | (1 2)(3 4) | 15 |
| $C_3$ | (3, 1, 1) | (1 2 3) | 20 |
| $C_4$ | (5) | (1 2 3) | 24 |

**Table 15.2** Conjugacy classes of $S_5$ contained in $A_5$.

these is $A_5$ which has order 60. It takes some thought to convince yourself that the numbers in the last column are correct. For example, for cycle type $(2, 2, 1)$, $(a\ b)(c\ d)$, there appear to be $5 \times 4 \times 3 \times 2 = 120$ choices for $a, b, c, d$. But $(a\ b) = (b\ a)$ and $(c\ d) = (d\ c)$. Also, disjoint cycles commute so $(a\ b)(c\ d) = (c\ d)(a\ b)$. Consequently, each element of this class is given by eight different choices of $a, b, c, d$. Hence, the class has $\frac{120}{8} = 15$ elements.

Each of these classes either remains a conjugacy class in $A_5$ or splits into two equal classes. Clearly, $C_1 = \text{conj}_{A_5}(\text{id})$. Consider $C_2 = \text{conj}_{S_5}((1\ 2)(3\ 4))$. The odd permutation $(1\ 2)$ commutes with $(1\ 2)(3\ 4)$ so, by Theorem 9(i), $\text{conj}_{A_5}((1\ 2)(3\ 4)) = C_2$. Hence, $C_2$ does not split into two. Similarly, $C_3$ does not split because $(1\ 2\ 3)$ commutes with the odd permutation $(4\ 5)$. However, $C_4$ must split into two because $|C_4| = 24$ is not a factor of $|A_5| = 60$. Thus, the class equation for $A_5$ is

$$60 = 1 + 15 + 20 + 12 + 12.$$

Suppose that $N$ is a normal subgroup of $A_5$. By Lagrange's Theorem, $|N|$ is a factor of 60. But $|N|$ must also be a sum of numbers in the list 1, 15, 20, 12, 12 of orders of conjugacy classes. Take a little time to convince yourself that it is not possible to write any factor of 60, except 1 and 60, as a sum of numbers in the list 1, 15, 20, 12, 12 with 1 included. This shows that $N$ must be the whole group $A_5$ or the trivial subgroup $\{\text{id}\}$ and establishes that $A_5$ is simple.

**EXERCISE 8**

Let $\alpha = (1\ 2\ 3\ 4)(5\ 6)$. Show that the conjugacy classes of $\alpha$ in $S_6$ and $A_6$ are equal.

**EXERCISE 9**

Let $\alpha = (1\ 2\ 3)(4\ 5\ 6)$. By rewriting $\alpha$ as $(4\ 5\ 6)(1\ 2\ 3)$, find an odd permutation $\theta$ in $S_6$ such that $\theta\alpha\theta^{-1} = \alpha$. Hence show that the conjugacy classes of $\alpha$ in $S_6$ and $A_6$ are equal.

## 15.7 Further Exercises on Chapter 15

**EXERCISE 10**

Let $m, n$ be integers $> 2$ such that $m$ is a factor of $n$, $n = dm$, say. Show that there is a homomorphism $f : D_n \to D_m$ with

$$f : \text{rot}_\theta \mapsto \text{rot}_{d\theta}; \quad f : \text{ref}_\theta \mapsto \text{ref}_{d\theta}.$$

What is the order of the kernel of $f$?

**EXERCISE 11**

For which values of $n$ does the dihedral group $D_n$ have a normal subgroup of order 2?

**EXERCISE 12**

Let $G_1$ and $G_2$ be the groups of symmetries of two equilateral triangles, each with centre at the origin. Show that there is a rotation $g \in O_2$ such that $G_2 = gG_1g^{-1}$.

## EXERCISE 13

Show that, in the dihedral group $D_n$, the cyclic subgroup $< \mathrm{rot}_{\frac{2\pi}{n}} >$ is normal.

## EXERCISE 14

Let $G$ be a group acting on a set $X$. Let $a \in G$, let $x \in X$ and let $H = \mathrm{stab}(x)$. Show that $\mathrm{stab}(a * x) = aHa^{-1}$.

## EXERCISE 15

Let $H$ be a subgroup of $G$ and let $g \in G$. Show that the left coset $gH$ of $H$ is a right coset of the conjugate $gHg^{-1}$.

## EXERCISE 16

Let $\alpha \in A_n$ be an even permutation which has a cycle of even length in its cycle decomposition. Show that the conjugacy classes of $\alpha$ in $S_n$ and $A_n$ are equal.

## EXERCISE 17

In $S_6$ there are six conjugacy classes consisting of even permutations.

(i) Write down one element from each class.

(ii) How many elements are in each class?

(iii) Which classes split into two conjugacy classes in $A_6$?

(iv) Show that the class equation for $A_6$ is

$$360 = 1 + 45 + 40 + 40 + 90 + 72 + 72.$$

(v) Check that, of the divisors of 360, only 1 and 360 can be written as a sum of numbers, including 1, in the list 1, 45, 40, 40, 90, 72, 72.

(vi) Show that $A_6$ is simple.

## EXERCISE 18

Let $\alpha = (1\ 2\ 3)(4\ 5\ 6\ 7\ 8)$.

(i) Write down the order of $< \alpha >$.

(ii) Compute the number of conjugates of $\alpha$ in $S_8$. Hence find the order of $\mathrm{cent}_{S_8}(\alpha)$; and show that $\mathrm{cent}_{S_8}(\alpha) =< \alpha >$.

(iii) Show that the conjugacy class of $\alpha$ in $S_8$ splits into two conjugacy classes in $A_8$.

## EXERCISE 19

Investigate when the conjugacy class of an even permutation in $S_n$ splits into two conjugacy classes in $A_n$. Can you suggest a criterion in terms of cycle decomposition for when this happens?

## EXERCISE 20

Let $G$ and $H$ be groups. What are the kernels of the projections $p_1 : G \times H \to G$ and $p_2 : G \times H \to H$?

## EXERCISE 21

In Chapter 13 we have seen that any group $G$ acts on itself by the rule $g * a = gag^{-1}$. Is this action always faithful? If not, what is the kernel?

## EXERCISE 22

Let $G, H$ be groups with direct product $G \times H$. Show that the subgroups $G' = \{(g, e_H) : g \in G\}$ and $H' = \{(e_G, h) : h \in H\}$ are normal subgroups of $G \times H$. Discuss whether the diagonal subgroup $D = \{(g, g) : g \in G\}$ is a normal subgroup of $G \times G$ (always?, never?, sometimes?, when?)

## EXERCISE 23

Use Lagrange's Theorem to show that a group of prime order must be simple.

## EXERCISE 24

Use Cauchy's Theorem to show that if $G$ is a finite abelian simple group then $G$ is cyclic of prime order.

# 16 • Factor Groups

The idea of a factor structure is fundamental in much of modern algebra. Our concern here is with factor groups that are groups formed from the cosets of a normal subgroup. After studying these through particular examples, we shall give an indication of their role in group theory by proving that, when $p$ is prime, all groups of order $p^2$ are abelian.

## 16.1 Cosets of the Kernel of an Action

Let $G$ be a group acting on the set $X$. In Chapter 11, we saw the significance of the left cosets of a stabilizer $H = \text{stab}(x)$. Any two elements of the same left coset $aH$ send $x$ to the same element $y = a * x$ of $X$. We now interpret the cosets of the kernel $K$ of an action in a similar way. The next theorem shows that all elements of $aK$ have the same effect on *every* element of $X$. In other words, all elements of the same coset perform the same permutation of $X$.

### • Theorem 1

Let $G$ be a group acting on a set $X$ and let $K$ be the kernel of the action. Let $a \in G$. Then the coset $aK$ is equal to the set $\{g \in G : g * x = a * x \text{ for all } x \in X\}$. ●

PROOF

Let $ak$ be any element of $aK$ and let $x \in X$. By the definition of kernel, $k * x = x$ so $(ak) * x = a * (k * x) = a * x$.

Conversely, suppose that $g * x = a * x$ for all $x \in X$. Let $x \in X$. Then

$$(a^{-1}g) * x = a^{-1} * (g * x) = a^{-1} * (a * x) = (a^{-1}a) * x = e * x = x$$

so $a^{-1}g \in K$. By Theorem 4(i) of Chapter 10, $gK = aK$ and hence $g \in aK$. ●

### Example 1

Let $G$ be the dihedral group $D_6$. We write the elements of $G$ as $e, r_i, 1 \leqslant i \leqslant 5$, $s_j, 1 \leqslant j \leqslant 6$, where $r_1 = \text{rot}_{\frac{\pi}{3}}$, $r_i = (r_1)^i$ and $s_j = \text{ref}_{\frac{(j-1)\pi}{3}}$. This group acts on the three diagonals of the hexagon, see Fig 16.1. The kernel $K$ of this action is the cyclic subgroup $< r_3 >$ generated by rotation through $\pi$. This rotation sends any line through the origin to itself. Since $|D_6| = 12$ and $|K| = 2$, there are six distinct cosets. These are shown in Fig 16.1 together with the corresponding permutations of the diagonals. In accordance with Theorem 1, the two elements of each coset perform the same permutation. The table shows that the six cosets of $K$ correspond to the six permutations which form the familiar group $S_3$. This suggests that the cosets of $K$ may form a group isomorphic to $S_3$.

To see what the binary operation in this possible group might be, consider the cosets $s_1K$ and $r_2K$. If you calculate $gh$ where $g$ is either element of $s_1K$ and $h$ is either element from $r_2K$, you obtain an element of the coset $s_5K$

$$s_1r_2 = s_5, \ s_1r_5 = s_2, \ s_4r_2 = s_2, \ s_4r_5 = s_5.$$

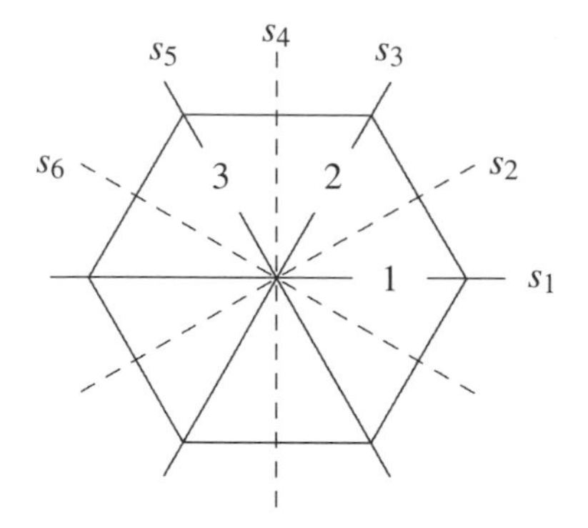


| Coset | Permutation |
|---|---|
| $K = \{e, r_3\}$ | id |
| $s_1K = \{s_1, s_4\} = s_4K$ | (2 3) |
| $s_2K = \{s_2, s_5\} = s_5K$ | (1 2) |
| $s_3K = \{s_3, s_6\} = s_6K$ | (1 3) |
| $r_1K = \{r_1, r_4\} = r_4K$ | (1 2 3) |
| $r_2K = \{r_2, r_5\} = r_5K$ | (1 3 2) |

**Fig 16.1** Action of $D_6$ on diagonals of a hexagon.

This calculation shows that $s_1Kr_2K = s_5K = (s_1r_2)K$, where $s_1Kr_2K$ is the product of the subsets $s_1K$ and $r_2K$ as defined on page 145.

We now aim to show that for the kernel $K$ of any group action, the cosets of $K$ form a group with binary operation $aKbK = (ab)K$.

## ● Theorem 2

Let $G$ be a group acting on a set $X$ and let $K$ be the kernel of the action. If $a, b \in G$ then the product $aKbK$ is equal to the left coset $(ab)K$. Thus, the set of cosets of $K$ in $G$ is closed under products. ●

PROOF

Let $ak_1bk_2 \in aKbK$ and let $x \in X$. Then

$$ak_1bk_2 * x = ak_1b * (k_2 * x) = ak_1b * x = a * (k_1 * (b * x)) = a * (b * x) = (ab) * x$$

because $k_1$ and $k_2$ are in the kernel of the action. By Theorem 1, $ak_1bk_2 \in (ab)K$ and hence $aKbK \subseteq (ab)K$. For the reverse inclusion, $abk = aebk \in aKbK$ for all $k \in K$. Therefore, $aKbK = (ab)K$. ●

## ● Theorem 3

Let $G$ be a group acting on a set $X$ and let $K$ be the kernel of the action. The set of cosets of $K$ in $G$ is a group with binary operation $aKbK = (ab)K$. ●

PROOF

We need to check the four group axioms for products of cosets of $K$.

**G1 [Closure]:** This is established in the above theorem, $aKbK = (ab)K$.

**G2 [Associativity]:** Using associativity in $G$

$$(aKbK)cK = (ab)KcK = ((ab)c)K = (a(bc))K = aK(bc)K = aK(bKcK).$$

**G3 [Neutral element]:** The obvious candidate for a neutral element is the left coset $eK$ arising from the neutral element $e$ in $G$. Recall from Theorem 4(ii) of Chapter 10 that

$eK = K$ so this particular coset is the subgroup $K$ itself. Let us check neutrality

$$eKaK = (ea)K = aK \text{ and } aKeK = (ae)K = aK \text{for all cosets } aK.$$

**G4 [Inverses]:** The obvious candidate for an inverse of $aK$ is the coset $a^{-1}K$ arising from the inverse of $a$ in $G$. Bearing in mind that our neutral element is $eK$, let us check that it is an inverse

$$aKa^{-1}K = (aa^{-1})K = eK \text{ and } a^{-1}KaK = (a^{-1}a)K = eK.$$ ●

The group given by this theorem is called the **factor group** of $G$ by $K$ and is written $G/K$.

## Example 2

Consider the dihedral group $D_4$ and its action on the diagonals of the square. Four elements of $D_4$, namely $e, r_2, s_2$ and $s_4$ stabilize both diagonals. These form the kernel $K$ of the action. The factor group $D_4/K$ has order 2 and its elements are $K$ and $r_1K$. The latter consists of the four elements which transpose or swop the two diagonals. Thus, $D_4$ is partitioned into two cosets which we might rename as

$$\text{stab} = \{e, r_2, s_2, s_4\} = K;$$

$$\text{swop} = \{r_1, r_3, s_1, s_3\} = r_1K.$$

The Cayley table for $D_4/K$ is as shown in Table 16.1. The names used here suggest a

| | stab | swop |
|---|---|---|
| stab | stab | swop |
| swop | swop | stab |

**Table 16.1** stab and swop.

group action of $D_4/K$ on the set $\{1, 2\}$ of diagonals:

$$\text{stab} * 1 = 1, \ \text{stab} * 2 = 2, \ \text{swop} * 1 = 2, \ \text{swop} * 2 = 1.$$

This is related to the action of $D_4$ on the diagonals but, unlike the action of $D_4$, it is faithful. Only the neutral element stabilizes the two diagonals.

We shall now see that for any action of a group $G$ on a non-empty set $X$ with kernel $K$, there is a corresponding faithful action of $G/K$ on $X$. Let $aK \in G/K$ and $x \in X$. By Theorem 1, all elements of $aK$ have the same effect as $a$ on $x$ so we can define $aK * x = a * x \in X$. We need to check that the rule $aK * x = a * x$ does define a group action.

**GA1:** The neutral element of $G/K$ is $eK$ and $eK * x = e * x = x$.

**GA2:** $(aKbK) * x = (abK) * x = (ab) * x = a * (b * x)$ and $aK * (bK * x) = aK * (b * x) = a * (b * x)$.

Thus, we do have an action of $G/K$ on $X$. This is sometimes called the **induced** action of $G/K$ on $X$. For the action of $G = D_6$ on the diagonals of the hexagon, the induced action

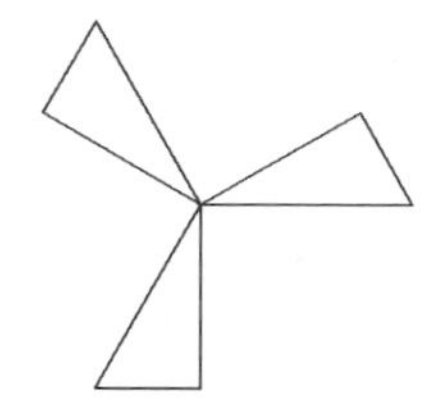

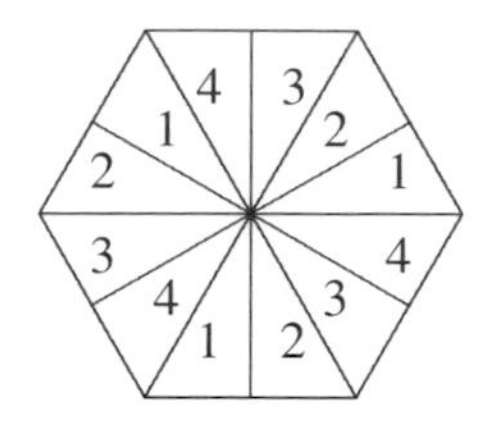


**Fig 16.2** A hexagon made from four fan shapes.

of the factor group $G/K$ is shown in Fig 16.1. It is faithful because the only coset performing the identity permutation on the diagonals is the kernel $K$ which is the neutral element of $G/K$.

● *Theorem 4*

Let $G$ be a group acting on a non-empty set $X$ and let $K$ be the kernel of the action. The induced action of $G/K$ on $X$ is faithful. ●

PROOF

We have to show that the only element $aK$ of $G/K$ for which $aK * x = x$ for all $x$ is the neutral element $eK$. But if $aK * x = x$ for all $x$ then, by definition of the induced action, $a * x = x$ for all $x$ that is, $a \in K$. But then, by Theorem 4(iii) of Chapter 10, $aK = K$. ●

**EXERCISE 1**

Figure 16.2 shows how a hexagon can be made from four fan shapes. The dihedral group $D_6$ acts on these fan shapes. For example $r_1 * 1 = 3$. Find the kernel $K$ of this action and draw up a table of the cosets of $K$ and the corresponding permutations as done in Fig 16.1. To which familiar group is the factor group $D_6/K$ isomorphic?

## 16.2 Factor Groups

Let $G$ be a group with a normal subgroup $N$. By Theorem 4(ii) of Chapter 15, $N$ is the kernel of an action of $G$. Hence, there is a factor group $G/N$ consisting of the cosets $aN$ of $N$ in $G$ under the binary operation $aNbN = (ab)N$.

There is a function $f : G \to G/N$ with $f(a) = aN$ for all $a \in G$. For $a, b \in G$ we have that $f(a)f(b) = aNbN = abN = f(ab)$ so $f$ is a homomorphism. What is its kernel? For $a \in G$, $f(a) \in \ker f$ if and only if $aN = eN$ so the kernel is $N$. Clearly $f$ is surjective. It is called the **natural homomorphism** from $G$ to $G/N$.

## Example 3

Let $G$ be the symmetric group $S_n$, $n \geqslant 2$, and let $N = A_n$ be the alternating group. As we saw in Chapter 15, $A_n$ is normal and its only cosets are $A_n$ itself and the set $B_n$ of all odd permutations in $S_n$. Thus, the factor group $S_n/A_n = \{A_n, B_n\}$ has order 2. Writing even for $A_n$ and odd for $B_n$, the Cayley table is on the left of Table 16.2.

| | even | odd |
|---|---|---|
| even | even | odd |
| odd | odd | even |

| | rot | ref |
|---|---|---|
| rot | rot | ref |
| ref | ref | rot |

**Table 16.2** even&odd and rot&ref.

## Example 4

We have seen in Chapter 15 that the special orthogonal or rotation group $SO_2$ is a normal subgroup of $O_2$ with just two cosets. One coset consists of all the rotations and the other consists of all the reflections. If we rename these as rot and ref, the factor group $O_2/SO_2$ has a Cayley table as shown on the right of Table 16.2. Note that here $G$ is infinite but, because there are only two distinct cosets, the factor group $G/N$ is finite. This group rot&ref also occurs as the factor group $G/N$, where $G = D_4$ is the group of symmetries of the square and $N$ is the subgroup of rotations of the square. This time the two elements are rot $= \{e, r_1, r_2, r_3\}$ and ref $= \{s_1, s_2, s_3, s_4\}$. The Cayley table for the factor group $G/N$ can be seen in that for the group $G = D_4$.

| $\mathbf{D_4}$ | $\mathbf{e}$ | $\mathbf{r_1}$ | $\mathbf{r_2}$ | $\mathbf{r_3}$ | $\mathbf{s_1}$ | $\mathbf{s_2}$ | $\mathbf{s_3}$ | $\mathbf{s_4}$ |
|---|---|---|---|---|---|---|---|---|
| $\mathbf{e}$ | $e$ | $r_1$ | $r_2$ | $r_3$ | $s_1$ | $s_2$ | $s_3$ | $s_4$ |
| $\mathbf{r_1}$ | $r_1$ | $r_2$ | $r_3$ | $e$ | $s_2$ | $s_3$ | $s_4$ | $s_1$ |
| $\mathbf{r_2}$ | $r_2$ | $r_3$ | $e$ | $r_1$ | $s_3$ | $s_4$ | $s_1$ | $s_2$ |
| $\mathbf{r_3}$ | $r_3$ | $e$ | $r_1$ | $r_2$ | $s_4$ | $s_1$ | $s_2$ | $s_3$ |
| $\mathbf{s_1}$ | $s_1$ | $s_4$ | $s_3$ | $s_2$ | $e$ | $r_3$ | $r_2$ | $r_1$ |
| $\mathbf{s_2}$ | $s_2$ | $s_1$ | $s_4$ | $s_3$ | $r_1$ | $e$ | $r_3$ | $r_2$ |
| $\mathbf{s_3}$ | $s_3$ | $s_2$ | $s_1$ | $s_4$ | $r_2$ | $r_1$ | $e$ | $r_3$ |
| $\mathbf{s_4}$ | $s_4$ | $s_3$ | $s_2$ | $s_1$ | $r_3$ | $r_2$ | $r_1$ | $e$ |

| | rot | ref |
|---|---|---|
| rot | rot | ref |
| ref | ref | rot |

## Theorem 5

The order of a factor group $G/N$ is the number of distinct cosets of $N$ in $G$. If $G$ is finite then

$$\left|\frac{G}{N}\right| = \frac{|G|}{|N|}.$$

PROOF

The first statement is immediate from the definition of $G/N$. The second follows from Lagrange's Theorem.

You may be wondering whether a factor group $G/H$ can be constructed when $H$ is a subgroup of $G$ that is not normal. The answer is no and the reason is that the formula $aHbH = (ab)H$ does not hold when $H$ is not normal. This is illustrated in the next example.

### Example 5

As in Example 1 of Chapter 10, let $G = D_4$ and $H =< s_4 >$. Then $r_1H = \{r_1, s_1\}$ and $r_3H = \{r_3, s_3\}$. Now $r_1r_3 = e$, $r_1s_3 = s_4$, $s_1r_3 = s_2$ and $s_1s_3 = r_2$ so $r_1Hs_2H$ has four distinct elements and cannot possibly be a left coset of $H$.

### Example 6

For a group $G$ in additive notation with a subgroup $N$, an arbitrary left coset of $N$ in $G$ is written $g + N$ rather than $gN$. We only use additive notation for abelian groups so such a subgroup $N$ is always normal. In additive notation, the binary operation in $G/N$ is

$$(g + N) + (k + N) = (g + k) + N.$$

We have already met factor groups of this type. If we take $G$ to be $\mathbb{Z}$ and $N$ to be the subgroup $n\mathbb{Z}$ of all multiples of some positive integer $n$ then each coset $g + N$ is the equivalence class $\overline{g}$ of $g$ under equivalence modulo $n$. The rule $(g + N) + (k + N) = (g + k) + N$ can then be rewritten $\overline{g} + \overline{k} = \overline{g + k}$. Thus, the factor group $\mathbb{Z}/n\mathbb{Z}$ is just the group $\mathbb{Z}_n$ under addition modulo $n$.

### Example 7

For every group $G$, the trivial subgroup $\{e\}$ and the whole group $G$ are normal subgroups of $G$ so we can form the factor groups $G/\{e\}$ and $G/G$. Neither of these is anything new. The latter only has one element as $G$ is the only coset of $G$ in $G$. If $N = \{e\}$ then each coset $gN$ consists of the single element $g$ and $G/N$ is just $G$ in disguise.

### EXERCISE 2

The alternating group $A_4$ has exactly one non-trivial proper normal subgroup $V$, see Exercise 7 of Chapter 15. Compute the order of the factor group $A_4/V$ and deduce that it is cyclic.

## 16.3 Calculations in Factor Groups

It is important when doing calculations in factor groups to remember that it is possible to have $aN = bN$ but $a \neq b$. In particular, it is possible to have $aN = N$ but $a \neq e$. We recall the rules for equality from Theorem 4 of Chapter 10

$$aN = bN \Leftrightarrow b^{-1}a \in N;$$

$$aN = N \Leftrightarrow a \in N.$$

## Example 8

Let $G$ be the dihedral group $D_4$ and let $N =< r_2 >$. Because $r_2$ is central in $D_4$, $N$ is normal in $G$. The factor group $G/N$ has order $\frac{8}{2} = 4$ and the four elements are

$$N = \{e, r_2\}; \quad r_1 N = \{r_1, r_3\};$$

$$s_1 N = \{s_1, s_3\}; \quad s_2 N = \{s_2, s_4\}.$$

To illustrate calculations in factor groups, we calculate $(r_1 N)^2$ and $s_1 N s_2 N$.

$(r_1 N)^2 = r_1 N r_1 N = r_2 N = N$ so $r_1 N$ has order 2.

$$s_1 N s_2 N = (s_1 s_2) N = r_3 N = r_1 N.$$

Notice that $s_2 N s_1 N = r_1 N = s_1 N s_2 N$ so that, although $s_1 s_2 \neq s_2 s_1$, the cosets $s_1 N$ and $s_2 N$ commute in the factor group.

### Order of elements in factor groups

Consider the factor group $G/K$ where $G = D_6$ and $K$ is as in Section 16.1. Table 16.1 exhibits an isomorphism between $G/K$ and $S_3$. Therefore, $G/K$ has three elements of order two and two elements of order three. The table indicates that the elements of order two in $G/K$ are cosets consisting of reflections. This suggests that perhaps the order of $aK$ and $a$ should be equal. However, the rotation $r_1$ has order six in $G$ whereas the coset $r_1 K$ has order three in $G/K$. The reason for this is that $r_1^3 = r_3 \in K$, so that $(r_1 K)^3 = (r_1)^3 K = r_3 K = K$. This illustrates the next theorem.

## Theorem 6

Let $G$ be a group with a normal subgroup $N$ and let $g \in G$. The coset $aN$ has finite order in $G/N$ if and only if $a^n \in N$ for some positive integer $n$. When this occurs, the order of $aN$ in $G/N$ is the least such positive integer. ●

PROOF

To decide whether $aN$ has finite order in $G/N$ and, if so, to compute the order, we need to determine when $(aN)^n$ is the neutral element $N$ of $G/N$. From the definition of the binary operation in $G/N$, we have the rule $(aN)^n = (a^n)N$ for all positive integers $n$. Hence, $(aN)^n = N$ precisely when $a^n \in N$. So $(aN)^n = e_{G/N}$ if and only if $a^n \in N$. The result follows immediately from this. ●

## Example 9

Let $G = C_9 =< g >$ be cyclic of order 9. As $G$ is abelian, every subgroup of $G$ is normal. We saw in Chapter 6 that $G$ has a unique subgroup $N$ of order 3, namely $< g^3 >$. So the factor group has order $\frac{9}{3} = 3$ and its elements are

$$N = \{e, g^3, g^6\} = g^3 N = g^6 N;$$

$$gN = \{g, g^4, g^7\} = g^4 N = g^7 N;$$

$$g^2 N = \{g^2, g^5, g^8\} = g^5 N = g^8 N.$$

We know from general theory that this factor group, being of order 3, must be cyclic with both $gN$ and $g^2N$ of order 3. As the next theorem shows, factor groups of cyclic groups are always cyclic.

## • Theorem 7

Let $G = < g >$ be a cyclic group and let $N$ be a subgroup of $G$. The factor group $G/N$ is cyclic generated by the coset $gN$. ●

PROOF

As $G$ is abelian, the factor group $G/N$ exists. Consider an arbitrary element $kN$ of $G/N$. As $G$ is generated by $g$, $k = g^j$ for some $j \in \mathbb{Z}$. Hence, $kN = g^j N = (gN)^j$ and so $G/N$ is cyclic generated by $gN$. ●

We consider now the orders of the factor groups $G/N$ where $G = C_n = < g >$ is cyclic of order $n$. We have seen in Chapter 6 that every subgroup $N$ of $G$ is cyclic generated by $g^m$ for some factor $m$ of $n$ and that $N$ then has order $d = \frac{n}{m}$. It follows from Theorems 7 and 5 that $G/N$ is cyclic of order $m$. With $d = \frac{n}{m}$, this can be summarized as saying that

$$\frac{C_n}{C_d} \cong C_m.$$

### EXERCISE 3

Let $G$, $N$ be as in Example 8. Then $G/N$ has order 4 and so is either cyclic or isomorphic to Klein's 4-group. Find the order of each of the three non-neutral elements of $G/N$ and decide whether $G/N$ is cyclic.

### EXERCISE 4

Let $G = C_{12}$ be cyclic of order 12, generated by $g$. Let $N$ be the cyclic subgroup generated by $g^6$.

(i) Write down the order of $G/N$.

(ii) Compute the order of $aN$ in $G/N$ when $a = g^4$ and when $a = g^3$.

## 16.4 The First Isomorphism Theorem

If $f : G \to H$ is an injective homomorphism then $G$ is isomorphic to a subgroup of $H$, namely the image $f(G)$. This is false if we drop the assumption of injectivity. For example, the action of $D_4$ on the two diagonals of the square gives rise to a homomorphism from $D_4$ to $S_2$ and it is impossible for a group of order 8 to be isomorphic to a subgroup of a group of order 2. In this section we show that, in the general case, the image $f(G)$ is isomorphic to a factor group of $G$, namely $G/K$ where $K$ is the kernel of $f$. Before we prove this in general, let us see that it holds for one familiar homomorphism.

### Example 10

The alternating group $A_n$ is the kernel of sgn : $S_n \to \mathbb{R}\backslash\{0\}$. The image of this homomorphism is $\{1, -1\}$ which is cyclic of order 2. The factor group $S_n/A_n$ is also cyclic of order 2 so the image is isomorphic to the factor group by the kernel.

In Example 4 of Chapter 15 we saw a homomorphism $f : D_4 \to S_4$ with kernel $\{e, r_2\}$ and observed that $f$ has the same effect on the two elements of any coset. For example, $f(s_1) = f(s_3)$. The next theorem shows that this is true in general.

### Theorem 8

Let $f : G \to H$ be a homomorphism with kernel $K$. Let $g \in G$ and let $a \in gK$. Then $f(a) = f(g)$.

PROOF

Let $a = gk \in gK$. Then $f(a) = f(gk) = f(g)f(k) = f(g)e_H = f(g)$.

Theorem 8 allows us to define a homomorphism from $G/K$ to $H$ whenever we have a homomorphism $f$ from $G$ to $H$ with kernel $K$. Let $gK \in G/K$. By Theorem 8, all elements of $gK$ are sent to $f(g)$ by $f$, so we can define a function $\overline{f} : G/K \to H$ by the rule

$$\overline{f}(gK) = f(g) \in H.$$

We need to check that $\overline{f}$ is a homomorphism. For this, let $gK, hK$ be any two elements of $G/K$. Then

$$\overline{f}(gKhK) = \overline{f}((gh)K) = f(gh) = f(g)f(h) = \overline{f}(gK)\overline{f}(hK).$$

Thus, $\overline{f} : G/K \to H$ is a homomorphism. It is sometimes called the **induced** homomorphism from $G/K$ to $H$.

Just as induced actions of factor groups are always faithful, induced homomorphisms from factor groups are always injective. To see this, it suffices, by Theorem 2 of Chapter 15, to show that the kernel of $\overline{f}$ is $\{e_{G/K}\}$. So let $gK \in G/K$ be such that $\overline{f}(gK) = e_H$, that is, $f(g) = e_H$. Then $g \in \ker f = K$ and $gK = K = e_{G/K}$. Hence, $\overline{f}$ is injective.

### Theorem 9 (The First Isomorphism Theorem)

Let $f : G \to H$ be a homomorphism with kernel $K$. The factor group $G/K$ is isomorphic to the image of $G$ under $f$, that is

$$G/\ker f \cong f(G).$$

PROOF

We first observe that the images $f(G)$ and $\overline{f}(G/K)$ of $f$ and $\overline{f}$ are equal. This is because $\overline{f}(gK) = f(g)$ for all $g \in G$. We have seen above that $\overline{f} : G/K \to H$ is injective so $G/K$ is isomorphic to the image $\overline{f}(G/K) = f(G)$.

### Example 11

Let $f : D_4 \to S_4$ be the homomorphism considered in Example 4 of Chapter 15. The kernel of $f$ is the cyclic subgroup $N =< r_2 >$, which is the centre of $D_4$, and we saw that the image $f(G)$ is a copy of Klein's 4-group $V$. Thus, $D_4/Z(D_4) \cong V$.

### Example 12

Consider the homomorphism $f$ from $\mathbb{R}$ (under addition) to the special orthogonal or rotation group $SO_2$ given by $f(\theta) = \text{rot}_\theta$ . The image is the whole group $SO_2$. The kernel $K$ consists of all integer multiples $2n\pi$ of $2\pi$, $n \in \mathbb{Z}$ and is the cyclic subgroup of $\mathbb{R}$ generated by $2\pi$. (Remember $\mathbb{R}$ is in additive notation.) Thus, $\mathbb{R}/K \cong SO_2$. This can be interpreted as saying that the rule for combining rotations in $SO_2$ really comes from addition of real numbers but with multiples of $2\pi$ 'factored out'.

When a group $G$ acts on a set $X$, there is a homomorphism $f : G \to S_X$ where, for $g \in G$, $f(g) = f_g$, the corresponding permutation of $X$. By definition, the kernel $K$ of the action is the kernel of $f$ so, by the First Isomorphism Theorem, $G/K \cong f(G)$. When $G = D_6$ acts on the diagonals of the hexagon, $f(G) = S_3$. The isomorphism between $G/K$ and $S_3$ is displayed in Fig 16.1.

### Example 13

Consider the action of the symmetric group $S_4$ on polynomials in four variables. One of the orbits for this action is the set $P = \{p_1, p_2, p_3\}$ where $p_1 = x_1x_2 + x_3x_4$, $p_2 = x_1x_3 + x_2x_4$ and $p_3 = x_1x_4 + x_2x_3$. The group $S_4$ acts on $P$ and gives rise to a homomorphism $f : S_4 \to S_3$. For example, the transposition $t = (1\ 2)$ stabilizes $p_1$ and interchanges $p_2$ and $p_3$. Hence, $f(t) = (2\ 3)$. The following display indicates that the image $f(S_4)$ is the whole of $S_3$

$$\begin{array}{rclcrcl} f((\text{id})) & = & \text{id} & \quad & f((1\ 2)) & = & (2\ 3) \\ f((2\ 3\ 4)) & = & (1\ 2\ 3) & & f((1\ 3)) & = & (1\ 3) \\ f((2\ 4\ 3)) & = & (1\ 3\ 2) & & f((1\ 4)) & = & (1\ 2). \end{array}$$

The kernel $K$ of this action is the normal subgroup $V = \{\text{id}, (1\ 2)(3\ 4), (1, 3)(2\ 4), (1\ 4)(2\ 3)\}$. Thus $S_4/V \cong S_3$.

### Example 14

Let $G$, $H$ be groups with direct product $G \times H$ and view $G$ and $H$ as subgroups of $G \times H$ as in Chapter 14. Consider the projection homomorphism $p_1 : G \times H \to G$, $(g, h) \mapsto g$. The kernel of $p_1$ is $H$ and $p_1$ is surjective so its image is $G$. Therefore, $\dfrac{G \times H}{H} \cong G$.

## EXERCISE 5

By considering a suitable homomorphism, show that the factor group $GL_2(\mathbb{R})/SL_2(\mathbb{R})$ of the general linear group by the special linear group is isomorphic to $(\mathbb{R}\backslash\{0\}, \times)$.

## 16.5 Groups of Order $p^2$ are Abelian

The purpose of this section is to illustrate the role of factor groups in group theory by proving that if $p$ is a prime number then all groups of order $p^2$ are abelian.

The basic idea in using factor groups $G/N$ is to put together information about the groups $N$ and $G/N$ to obtain information about $G$. We shall do this in the case where $N = Z(G)$ is the centre of $G$, as defined in Chapter 13. The group $Z(G)$ is abelian because any element of $Z(G)$ commutes with all elements of $G$ and hence with all elements of $Z(G)$. Moreover, $Z(G)$ is always normal, because $gng^{-1} = ngg^{-1} = n$ for all $g \in G$ and all $n \in Z(G)$. Hence, we can form the factor group $G/Z(G)$.

## Theorem 10

If $G$ is a group such that $G/Z(G)$ is cyclic then $G$ is abelian.

PROOF

Let $Z = Z(G)$ and let $gZ$ be a generator of $G/Z$. Then every element of $G/Z$, that is every coset of $Z$, has the form $(gZ)^m = (g^m)Z$, $m \in \mathbb{Z}$. But the cosets of $Z$ give a partition of $G$ so each element $a$ of $G$ is in one of these. If $a \in (g^m)Z$ then $a = g^m z$ for some $z \in \mathbb{Z}$. Now let $a, b \in G$. Then $a = g^{m_1} z_1$ and $b = g^{m_2} z_2$ for some $m_1, m_2 \in \mathbb{Z}$ and some $z_1, z_2 \in Z$. Since $z_1$ and $z_2$ commute with every element of $G$ and powers of $g$ commute with each other

$$ab = g^{m_1+m_2} z_1 z_2 = ba.$$

Thus, $G$ is abelian.

## Theorem 11

If $p$ is a prime number then every group $G$ of order $p^2$ is abelian.

PROOF

Let $Z$ be the centre of $G$. We have to show that $Z = G$. By Lagrange's Theorem, $|Z|$ must be one of $1, p, p^2$, the only factors of $p^2$. By Theorem 5 of Chapter 13, the centre is non-trivial, that is $|Z| \neq 1$. Suppose that $|Z| = p$. The factor group $G/Z$ has order $\frac{p^2}{p} = p$. By Theorem 8 of Chapter 10, all groups of prime order are cyclic so $G/Z$ is cyclic. Theorem 10 then says that $G$ must be abelian, that is $|Z| = |G| = p^2$. Thus, the possibility $|Z| = p$ does not arise. The only remaining possibility is that $|Z| = p^2$ and so $Z = G$. Hence, $G$ is abelian.

## Example 15

One of our main examples of groups, $D_4$, has order $8 = 2^3$ and is not abelian so the above theorem cannot be extended to higher powers of $p$ than the square. Of course $D_4$ has a non-trivial centre $Z =< r_2 >$ and we have seen that $D_4/Z$ is isomorphic to Klein's 4-group.

## EXERCISE 6

Let $G$ be a group of order 91 with a non-trivial centre. Show that $G$ is abelian.

## EXERCISE 7

Give an example of a group $G$ with a normal subgroup $N$ such that both $N$ and $G/N$ are cyclic but $G$ is not abelian.

## 16.6 Further Exercises on Chapter 16

### EXERCISE 8

In the factor group $S_4/V$ discussed in Example 13, find the order of the coset $aV$ when $a = (1\,2\,3\,4)$.

### EXERCISE 9

The vertices of a regular hexagon are the vertices of two equilateral triangles, see Fig 14.1. The group $D_6$ acts on these two triangles. Find the kernel $K$ of the action and show that $D_6/K \cong C_2$.

### EXERCISE 10

By considering various actions of the dihedral group $D_{12}$, show that the dihedral groups $D_6$, $D_3$, and $D_4$, Klein's 4-group $V$ and the cyclic group $C_2$ are all isomorphic to factor groups of $D_{12}$.

### EXERCISE 11

Show that every factor group of an abelian group is abelian. Is it true that every factor group of a non-abelian group is non-abelian?

### EXERCISE 12

Show that for all positive integers $m, n$ with $n \geqslant 3$, the dihedral group $D_n$ is isomorphic to a factor group of $D_{mn}$.

### EXERCISE 13

The set of all invertible upper triangular $2 \times 2$ real matrices $\begin{pmatrix} a & b \\ 0 & d \end{pmatrix}$, where $ad \neq 0$, is a subgroup $G$ of $GL_2(\mathbb{R})$. Let $K = \left\{\begin{pmatrix} 1 & b \\ 0 & 1 \end{pmatrix}\right\}$ be the shear group and let $H = (\mathbb{R}\backslash\{0\}, \times)$. By considering the function $f : G \to H \times H$ where $f : \begin{pmatrix} a & b \\ 0 & d \end{pmatrix} \mapsto (a, d)$ show that $K$ is a normal subgroup of $G$ and that $G/K \cong H \times H$.

## EXERCISE 14

Let $G$ and $H$ be groups with normal subgroups $N$ and $M$ respectively. Show that $N \times M$ is a normal subgroup of $G \times H$ and that

$$\frac{G \times H}{N \times M} \cong \frac{G}{N} \times \frac{H}{M}.$$

## EXERCISE 15

(This exercise shows that it is possible to have two isomorphic normal subgroups $N_1$ and $N_2$ of a group $G$ with non-isomorphic factor groups.) Let $G = S_3 \times C_3$ and let $c$ be a generator of $C_3$.

(i) Let $N_1$ be the cyclic subgroup generated by $((1\ 2\ 3), e)$. Show that $G/N_1 \cong C_6$.

(ii) Let $N_2$ be the cyclic subgroup generated by $(\mathrm{id}, c)$. Show that $G/N_2 \cong S_3$.

## EXERCISE 16

Show that $S_5$ has no normal subgroup of order 5 and hence that $S_4$ is not isomorphic to a factor group of $S_5$.

## EXERCISE 17

Let $n \geqslant 3$. Show that Klein's 4-group is isomorphic to a factor group of the dihedral group $D_n$ if and only if $n$ is even.

## EXERCISE 18

Let $G$ be an abelian group and let $D$ be the diagonal subgroup of $G \times G$. By considering an appropriate homomorphism $f : G \times G \to G$, show that $D$ is a normal subgroup of $G \times G$ and that $\dfrac{G \times G}{D} \cong G$.

## EXERCISE 19

Let $G$ be a group of order 102 with central elements of orders 2 and 3. Show that $G$ is abelian.

# 17 • Groups of Small Order

In this chapter we classify, up to isomorphism, groups of certain orders including those up to 12. In other words, for these orders, we give a list of groups of the appropriate order such that every group of that order is isomorphic to one of the groups in the list. The theorems which we shall use in classifying groups of small order are

- from Chapter 10, Lagrange's Theorem and its consequences that the order of any element of a finite group $G$ divides $|G|$ and that every group of prime order is cyclic;
- from Chapter 14, Cauchy's Theorem and Theorem 2 on recognizing direct products;
- from Chapter 16, for each prime number $p$, every group of order $p^2$ is abelian.

## 17.1 Groups of Order $p^2$

Exercise 5 of Chapter 10 classified groups of order 4. Any group of order 4 is isomorphic either to the cyclic group $C_4$ or to the direct product $C_2 \times C_2$. This is typical of groups of order $p^2$ where $p$ is prime.

Let $p$ be a prime number. We know that all groups of order $p^2$ are abelian. The two groups of order $p^2$ which we have met are the cyclic group $C_{p^2}$ and the direct product $C_p \times C_p$, which, by Theorem 9 of Chapter 6, is not cyclic. As in the case $p = 2$, these turn out to be the only groups of order $p^2$. In particular there are only two non-isomorphic groups of order 9.

● *Theorem 1*

Every group of order $p^2$ is isomorphic to either $C_{p^2}$ or $C_p \times C_p$. ●

PROOF

Let $G$ be a group of order $p^2$ and suppose that $G$ is not cyclic. We shall show that $G \cong C_p \times C_p$. As $G$ is not cyclic, it has no element of order $p^2$. Only the neutral element has order 1 so $G$ has $p^2 - 1$ elements of order $p$. If $a$ is any one of these elements then it generates a cyclic subgroup $H$ of order $p$. As $|H| < |G|$ there exists $b \in G$ with $b \notin H$. Let $K = < b >$ which also has order $p$. By Theorem 9 of Chapter 10, $H \cap K = \{e\}$. As $G$ is abelian, it follows from Theorem 2 of Chapter 14 that $HK$ is a subgroup of $G$ of order $p^2$, isomorphic to $C_p \times C_p$. As $|HK| = |G|$ we must have that $G = HK \cong C_p \times C_p$. ●

## 17.2 Groups of Order $2p$

Two of the numbers $\leqslant 12$ for which we do not yet have a classification are 6 and 10. These both have the form $2p$ where $p$ is an odd prime. Groups of order $2p$ which we have met are the cyclic group $C_{2p}$, the direct product $C_p \times C_2$ and the dihedral group $D_p$. The first two of these are isomorphic by Theorem 9 of Chapter 6 so we just have $C_{2p}$ and $D_p$. In the case $p = 3$, the symmetric group $S_3$ has order 6 but we saw in Chapter 9 that

$S_3 \cong D_3$. The aim of this section is to show that any non-cyclic group of order $2p$ is isomorphic to the dihedral group $D_p$. First we look at the basic properties of the dihedral groups, $D_n$.

In $D_n$, let $a = \text{rot}_{\frac{2\pi}{n}}$ and $b = \text{ref}_0$. Then

$$ab = \text{rot}_{\frac{2\pi}{n}} \text{ref}_0 = \text{ref}_{\frac{2\pi}{n}} = \text{ref}_0 \text{rot}_{-\frac{2\pi}{n}} = ba^{-1} = ba^{n-1}.$$

Every element of $D_n$ has one of the forms $a^m = \text{rot}_{\frac{2m\pi}{n}}$ or $ba^m = \text{ref}_{\frac{-2m\pi}{n}}$, where $0 \leqslant m < n$. In order to express the product of any two of these elements in one of these forms, it is enough to know that $ab = ba^{n-1}$. For example, when $n = 5$

$$a^2(ba^3) = a(ab)a^3 = aba^4a^3 = ba^4a^4a^3 = ba^{11} = ba.$$

In general, $a^m ba^q = ba^{m(n-1)+q}$.

## Theorem 2

Let $G$ be a group of even order $2n$. Suppose that $G$ has an element $a$ of order $n$ and an element $b$ of order 2 with $b \notin < a >$.

(i) $bab^{-1} = a^j$ for some integer $j$ such that $0 < j < n$ and $n$ is a factor of $j^2 - 1$.

(ii) If $n \geqslant 3$ and $bab^{-1} = a^{n-1}$ then $G \cong D_n$.

PROOF

(i) Let $H = < a >$ which has order $n$ and index 2 in $G$. By Theorem 8 of Chapter 15, $H$ is normal. Hence $bab^{-1} \in H$, that is, $bab^{-1} = a^j$ for some $j$ with $0 \leqslant j < n$. Clearly, $j \neq 0$ for otherwise $ba = b$ and $a = e$. Now $b$ has order 2 so $b^2 = b^{-2} = e$. Therefore

$$\begin{aligned} a = b^2ab^{-2} &= b(bab^{-1})b^{-1} \\ &= b(a^j)b^{-1} \\ &= (bab^{-1})^j \text{ (because conjugation by } b \text{ is a homomorphism)} \\ &= (a^j)^j = a^{(j^2)} \end{aligned}$$

As $a$ has order $n$ and $a = a^{j^2}$, it follows from Theorem 3(iv) of Chapter 6 that $n$ divides $j^2 - 1$.

(ii) As $H$ has index 2 in $G$, it only has two cosets in $G$. One of these is $H$ and, because $b \notin H$, the other must be $bH$. Each element of $G$ must be in $H$ or $bH$ and so must be of the form $a^m$ or of the form $ba^m$ for some $m$ with $0 \leqslant m < n$. Because $bab^{-1} = a^{n-1}$ and $b^{-1} = b$, we have that $ab = ba^{n-1}$. As for the dihedral group, we can use the formula $ab = ba^{n-1}$ to express the product of any two elements in one of the forms $a^m$ or $ba^m$. It follows that the bijective function $f : G \to D_n$ with

$$f(a^m) = \text{rot}_{\frac{2m\pi}{n}}, \text{ and } f(bo^m) = \text{ref}_{\frac{-2m\pi}{n}}$$

is an isomorphism.

● *Theorem 3*

If $p$ is an odd prime then every group of order $2p$ is isomorphic to either $C_{2p}$ or $D_p$. ●

PROOF

Let $G$ be a non-cyclic group of order $2p$. By Cauchy's Theorem, $G$ has an element $a$ of order $p$ and an element $b$ of order 2. Then $b \notin < a >$ because its order, 2, is not a factor of $p = | < a > |$.

By Theorem 2(i), $bab^{-1} = a^j$ where $0 < j < p$ and $p$ divides $j^2 - 1 = (j+1)(j-1)$. As $p$ is prime, either $p$ divides $j + 1$ or $p$ divides $j - 1$. Because $1 \leqslant j \leqslant p - 1$, the only possibilities are $j = p - 1$ and $j = 1$.

If $j = 1$ then $ab = ba$ and, by Theorem 10 of Chapter 10, $ab$ has order $2p$. In this case, $G = < ab > \cong C_{2p}$ is cyclic.

If $j = p - 1$ then, by Theorem 2(ii), $G \cong D_p$. ●

Theorem 3 shows that all groups of order 6 are isomorphic to either $C_6$ or $D_3$ and that all groups of order 10 are isomorphic to either $C_{10}$ or $D_5$.

**EXERCISE 1**

Let $G$ be a non-cyclic group of order 10 and let $a$ and $b$ be elements of orders 5 and 2 respectively. (Thus, $bab^{-1} = a^4$ by the proof of Theorem 3.) Express each of the elements $a^3b$ and $ba^2b$ in one of the forms $a^m$, $ba^m$ with $0 \leqslant m < 5$. Confirm your answers by setting $a = \mathrm{rot}_{\frac{2\pi}{5}}$, $b = \mathrm{ref}_0$ in $D_5$ and computing products using Table 1.1.

## 17.3 Groups of Order 8

The groups of order 8 which we have seen are the cyclic group $C_8$, the direct products $C_4 \times C_2$ and $C_2 \times C_2 \times C_2$ and the dihedral group $D_4$. We shall see that this list is not yet complete.

Let $G$ be a group of order 8. The possible orders of elements of $G$ are 1,2,4 and 8. If $g$ has order 4 or 8 then either $g^2$ or $g^4$ has order 2 so $G$ has at least one element of order 2. Alternatively, Cauchy's Theorem gives the same conclusion. We now consider several cases.

### Case 1

If $G$ has an element $g$ of order 8 then $G = < g >$ is cyclic, $G \cong C_8$.

### Case 2

Suppose that, apart from $e$, every element of $G$ has order 2. Then every element of $G$ is its own inverse and, for $a, b \in G$

$$ab = (ab)^{-1} = b^{-1}a^{-1} = ba,$$

whence $G$ is abelian. If $a \neq e$ and $b \notin < a >$ then we obtain a subgroup $H = \{e, a, b, ab\}$ isomorphic to $C_2 \times C_2$, Klein's 4-group. With $K = < c >$, where $c$ is any of the four elements of $G \backslash H$, Theorem 2 of Chapter 14 applies to show that

$G = HK \cong C_2 \times C_2 \times C_2$. An example of this case is the group $U(\mathbb{Z}_{24})$ which was analysed in Example 4 of Chapter 14.

## Case 3

Suppose that $G$ has an element $a$ of order 4 and an element $b$ of order 2 such that $b \notin \langle a \rangle$. By Theorem 2(i), $bab^{-1} = a^j$ for some $j$ such that $0 < j < 4$ and 4 divides $j^2 - 1$. The only possibilities are $j = 1$ and $j = 3$. If $j = 3$ then, by Theorem 2(ii), $G \cong D_4$. If $j = 1$ then $ba = ab$ and, setting $K = \langle b \rangle$ and, applying Theorem 2 of Chapter 14, $G = HK \cong C_4 \times C_2$.

## Case 4

The cases covered so far include all the groups listed at the beginning of the section. Can there be any other groups of order 8?

### Theorem 4

Let $G$ be a group of order 8 which is not isomorphic to any of the groups $C_8, C_4 \times C_2, C_2 \times C_2 \times C_2, D_4$. Then $G$ has exactly one element $d$ of order 2 and six elements, $a^{\pm 1}, b^{\pm 1}$ and $c^{\pm 1}$, of order 4. Also $G = \{e, a, a^2, a^3, b, ba, ba^2, ba^3\}$ and $ab = ba^3$.

PROOF

As $G \not\cong C_8$, it has no element of order 8. Moreover, $G$ must have at least one element, $a$ say, of order 4, otherwise Case 2 would apply to give $G \cong C_2 \times C_2 \times C_2$. Let $d = a^2$ and let $g$ be any element of order 2. Then $g \in \langle a \rangle$, otherwise we would have the situation of Case 3 and $G \cong D_4$ or $G \cong C_4 \times C_2$. The only element of $\langle a \rangle$ of order 2 is $d$ so $g = d$. Hence the only element of $G$ of order 2 is $d$ and there are six elements of order 4, which must occur as three inverse pairs, $a^{\pm 1}, b^{\pm 1}$ and $c^{\pm 1}$.

Let $H = \langle a \rangle = \{e, a, a^2, a^3\}$. As $H$ has index 2 in $G$, it only has two cosets in $G$. One of these is $H$ and, because $b \notin H$, the other must be $bH = \{b, ba, ba^2, ba^3\}$. Hence, the eight distinct elements of $G$ are as stated. Notice that $d = a^2$, $b^{-1} = b^3 = bb^2 = bd = ba^2$ and $c$ must be either $ba$ or $ba^3$.

Since $ab \neq e$ and $ab \neq d$, the element $ab$ has order 4. The square of each element of order 4 has order 2 and so must be $d$. Hence, $(ab)^2 = d = b^2$, that is, $abab = bb$. By cancellation, $aba = b$ and so $ab = ba^{-1} = ba^3$.

Theorem 4 shows that, up to isomorphism, there is at most one group of order 8 not in our original list. If such a group exists then its elements are as in Theorem 4. Now $a^4 = e = b^4$ and $ab = ba^3$. These rules would then determine all the products $a^i a^j$, $ba^i a^j$, $a^i ba^j$ and $ba^i ba^j$ in $G$, for example $a^2 b = aba^3 = ba^6 = ba^2$ and $a^3 b = a(a^2 b) = aba^2 = ba^5 = ba$. But does such a group $G$ exist?

The missing group does exist and is called the quaternion group $Q$. It lives inside a larger algebraic structure, the so-called division ring of quaternions, which was discovered by **Hamilton**. This consists of elements of the form $\alpha + \beta i + \gamma j + \delta k$, where $\alpha, \beta, \gamma, \delta \in \mathbb{R}$. These can be added together to give a group under addition, isomorphic to $\mathbb{R} \times \mathbb{R} \times \mathbb{R} \times \mathbb{R}$. More interestingly, they can be multiplied together, using the rules

$$i^2 = j^2 = k^2 = -1; \quad ij = k = -ji, \; jk = i = -kj, \; ki = j = -ik.$$

The non-zero elements then form an infinite non-abelian group with $Q = \{\pm 1, \pm i, \pm j, \pm k\}$ as a subgroup of order 8. The Cayley table is shown in Table 17.1.

| $Q$ | $1$ | $i$ | $-1$ | $-i$ | $j$ | $k$ | $-j$ | $-k$ |
|---|---|---|---|---|---|---|---|---|
| $1$ | $1$ | $i$ | $-1$ | $-i$ | $j$ | $k$ | $-j$ | $-k$ |
| $i$ | $i$ | $-1$ | $-i$ | $1$ | $k$ | $-j$ | $-k$ | $j$ |
| $-1$ | $-1$ | $-i$ | $1$ | $i$ | $-j$ | $-k$ | $j$ | $k$ |
| $-i$ | $-i$ | $1$ | $i$ | $-1$ | $-k$ | $j$ | $k$ | $-j$ |
| $j$ | $j$ | $-k$ | $-j$ | $k$ | $-1$ | $i$ | $1$ | $-i$ |
| $k$ | $k$ | $j$ | $-k$ | $-j$ | $-i$ | $-1$ | $i$ | $1$ |
| $-j$ | $-j$ | $k$ | $j$ | $-k$ | $1$ | $-i$ | $-1$ | $i$ |
| $-k$ | $-k$ | $-j$ | $k$ | $j$ | $i$ | $1$ | $-i$ | $-1$ |

**Table 17.1** Cayley table for $Q$.

This fits the description given in Theorem 4 with $a = i$ and $b = j$. The unique element of order 2 is $-1$ and each of the other six non-neutral elements has order 4.

We have not actually shown that $Q$ is a group as we have not checked for associativity. Another disadvantage of this approach is that checking for closure and inverses involves computation of the Cayley table. For larger groups this becomes impractical.

An alternative approach to the construction of $Q$ is to exhibit it as a stabilizer for a group action. There is then no further need to check for closure, associativity or inverses as stabilizers are always subgroups. As we have already done for the alternating group $A_n$, we now construct $Q$ as the stabilizer of a polynomial in several variables. A similar method will be used in the next section to construct a group of order 12 which we have not yet met.

Let $p$ be the polynomial

$$x_1x_2^2x_5^2x_6^3 + x_2x_3^2x_6^2x_7^3 + x_3x_4^2x_7^2x_8^3 + x_4x_1^2x_8^2x_5^3$$
$$+x_5x_8^2x_3^2x_2^3 + x_6x_5^2x_4^2x_3^3 + x_7x_6^2x_1^2x_4^3 + x_8x_7^2x_2^2x_1^3.$$

Let $Q = \text{stab}(p)$ for the action of the symmetric group $S_8$ on polynomials in eight variables. If $a = (1\ 5\ 3\ 7)(8\ 4\ 6\ 2)$ and $b = (1\ 2\ 3\ 4)(5\ 6\ 7\ 8)$ then you can check that

- $a * p = p = b * p$ so $a \in Q$ and $b \in Q$;
- $a^4 = b^4 = \text{id}$ and $a^2 = b^2 = (1\ 3)(2\ 4)(5\ 7)(6\ 8) = d$, say;
- $a^{-1} = (7\ 3\ 5\ 1)(2\ 6\ 4\ 8) = a^3$ and $b^{-1} = (4\ 3\ 2\ 1)(8\ 7\ 6\ 5) = b^3$;
- $ab = (1\ 8\ 3\ 6)(2\ 7\ 4\ 5) = ba^3$ and $ba = (1\ 6\ 3\ 8)(2\ 5\ 4\ 7)$.

Thus, $Q$ contains the eight elements $e, a, a^2 = d, a^3 = a^{-1}, b, ba, ba^2$ and $ba^3$ and these fit the description in Theorem 4.

Now $|Q|$ is at least 8 but could it be greater than 8? Any permutation which stabilizes $p$ must permute the eight terms $t_1 = x_1x_2^2x_5^2x_6^3, t_2, \ldots, t_8$ whose sum is $p$. For example,

$a * t_1 = t_2$. Thus, $Q$ acts on $\{t_1, t_2, \ldots, t_8\}$. Applying each of the eight elements of $Q$ listed above to $t_1$, we see that the orbit of $t_1$ contains all eight terms. By the finite form of the Orbit-Stabilizer Theorem, $|Q| = |\operatorname{orb} t_1||\operatorname{stab}(t_1)| = 8 \times |\operatorname{stab}(t_1)|$, where $\operatorname{stab}(t_1)$ is the stabilizer of $t_1$ in $Q$. To show that $|Q| = 8$, we need to check that $\operatorname{stab}(t_1)$ has order 1.

Let $g \in \operatorname{stab}(t_1)$. Then $g * p = p$ (otherwise $g \notin Q$) and $g * t_1 = t_1$. As $t_1 = x_1 x_2^2 x_5^2 x_6^3$ is stabilized, $g(1) = 1$ and $g(6) = 6$. Hence, $g * t_7 = t_7$, as this is the only term featuring $x_1^2 x_6^2$, and so $g(7) = 7$ and $g(4) = 4$. Repeating the argument, $g * t_3 = t_3$, $g(3) = 3$, $g(8) = 8$, $g * t_5 = t_5$, $g(5) = 5$ and $g(2) = 2$. Thus, $g = \text{id}$. Therefore, $|\operatorname{stab}(t_1)| = 1$ and $|Q| = 8$.

We now have constructed the missing group of order 8 in two ways. Theorem 4 together with the case-by-case analysis that preceded it gives the following classification theorem for groups of order 8.

● *Theorem 5*

Every group $G$ of order 8 is isomorphic to one of the groups
$C_8, C_4 \times C_2, C_2 \times C_2 \times C_2, D_4, Q$. ●

**EXERCISE 2**

Find the centre of the quaternion group $Q$, the centralizer of $i$ in $Q$, the conjugacy class of $i$ in $Q$ and the class equation for $Q$.

## 17.4 Groups of Order 12

Groups of order 12 include the cyclic group $C_{12}$, the direct product $C_6 \times C_2$, the dihedral group $D_6$ and the alternating group $A_4$. The direct products $C_3 \times C_4$ and $C_3 \times C_2 \times C_2$ are, by Theorem 9, isomorphic to $C_{12}$ and $C_6 \times C_2$ respectively. Also, we have seen in Chapter 14 that $D_3 \times C_2 \cong D_6$. As in our study of groups of order 8, there is one group of order 12 missing from this list and it will be constructed as the stabilizer of a polynomial.

● *Theorem 6*

Let $G$ be an abelian group of order 12. Then either $G \cong C_6 \times C_2$ or $G \cong C_{12}$. ●

PROOF

By Cauchy's Theorem, $G$ has an element $a$ of order 2 and an element $b$ of order 3. As these commute, their product $c = ab$ has order 6. Let $d \in G$ with $d \notin < c >$. Then $d$ has order 2,3,4,6 or 12. If $d$ has order 2 then $G = HK \cong C_6 \times C_2$ where $H = < c >$ and $K = < d >$. If $d$ has order 3 then $G$ would have a subgroup $< b >< d >$ of order 9 which is impossible by Lagrange's Theorem. If $d$ has order 4 then $bd$ has order 12 and $G = < bd >$ is cyclic, that is $G \cong C_{12}$. If $d$ has order 12 then $G = < d > \cong C_{12}$. We are left with the possibility that all elements $d$ of $G$ not in $< c >$ have order 6. If $d$ has order 6 then $d^2$ and $d^3$ have orders 3 and 2 respectively, and so must be in $< c >$. Hence $d = d^3 d^{-2} \in < c >$, contradicting the choice of $d$. Thus, if $G$ is abelian of order 12, either $G \cong C_6 \times C_2$ or $G \cong C_{12}$. ●

## • Theorem 7

Let $G$ be a non-abelian group of order 12 with an element $a$ of order 6 and an element $b$ of order 2 with $b \notin < a >$. Then $G \cong D_6$. •

PROOF

By Theorem 2(i), $bab^{-1} = a^j$ for some $j$ such that $0 < j < 6$, and 6 divides $j^2 - 1$. The only possibilities are $j = 1$ and $j = 5$. If $ba = ab$ then $G = < a >< b > \cong C_6 \times C_2$ which is abelian. Hence, $j = 5$ and, by Theorem 2(ii), $G \cong D_6$. •

## • Theorem 8

Let $G$ be a group of order 12.

(i) If $G$ has a subgroup $H$ of order 3 which is not normal then $G \cong A_4$.

(ii) If $G$ has no element of order 6 then $G \cong A_4$. •

PROOF

(i) We use the action of $G$ on the set $X$ of the four left cosets of $H$ in $G$ introduced in Section 15.3. Thus, $g * aH = (ga)H$. The kernel $K$ of this action is contained in $H = \text{stab}(H)$ but is normal in $G$. As $H$, being of prime order, has only two subgroups, $H$ and $\{e\}$, and is not itself normal in $G$, $K = \{e\}$. Thus, the action is faithful. Hence $G$ is isomorphic to the image $f(G)$ where $f : G \to S_4$ is the homomorphism arising from this action. We have seen in Example 10 of Chapter 15 that $S_4$ only has one subgroup of order 12, namely $A_4$. Therefore, $f(G) = A_4$ and $G \cong A_4$.

(ii) By Cauchy's Theorem, $G$ has an element $a$ of order 3. Let $H =< a >$. Suppose that $H$ is normal. Then $bab^{-1} \in H = \{e, a, a^2\}$ for all $b \in G$. The possibility $bab^{-1} = e$ cannot occur because $a \neq e$ so $bab^{-1} = a$ or $a^2$.

The centralizer $\text{cent}(a)$ of $a$ in $G$ is a subgroup of $G$ containing $H$ so, by Lagrange's Theorem, it has order 3,6 or 12. If it has order 6 or 12 then, by Cauchy's Theorem it has an element $g$, say, of order 2. As $ag = ga$, the element $ag$ would have order 6. As $G$ has no such element, $|\text{cent}(a)| = 3$ and hence $\text{cent}(a) = H$.

Let $b, c \in G \backslash H$. Then $b \notin \text{cent}(a)$ so $ba \neq ab$ and hence $bab^{-1} \neq a$. Therefore, $bab^{-1} = a^2$ and $ba = a^2b$. Similarly, $ca = a^2c$. Now consider $bc$. Then

$$bca = ba^2c = (ba)ac = a^2(ba)c = a^2a^2bc = a^4bc = abc.$$

Thus, $bc \in \text{cent}(a) = H$ for all $b, c \in G \backslash H$. But for each $b \in G \backslash H$, the nine different elements $c \in G \backslash H$ give rise, by the cancellation laws, to nine different elements $bc$. These cannot all be in $H$. Hence, $H$ cannot be normal and so, by (i), $G \cong A_4$. •

The only possibility not covered by these three theorems is where $G$ is not abelian, and $G$ has an element $a$ of order 6 such that every element of order 2 in $G$ is in $< a >$. So $G$ has a unique element $d = a^3$ of order 2. This is reminiscent of the situation which led us to construct the quaternion group $Q$ in the previous section. Let $H = \langle a \rangle$ and note that, in $H$, the orders of the elements $e, a, a^2, a^3 = d, a^4$ and $a^5 = a^{-1}$ are 1, 6, 3, 2, 3 and 6 respectively.

We shall show that the six elements of $G$ not in $H$ must all have order 4. To see this, let $b \in G$ with $b \notin H$. Then $b$ has order 3,4 or 6 because the only element of order 2 is in $H$ and, as $G$ is not abelian, there can be no element of order 12. As $H$ has index 2 it is normal in $G$. Moreover, $H$ and $bH$ are the only cosets of $H$ in $G$ so every element of $G$ has one of the forms $a^i$, $ba^i$. As $G$ is not abelian, $ba \neq ab$, otherwise all pairs of elements commute. Because $H$ is normal, $bab^{-1} \in H$ and, because $ba \neq ab$, it cannot be $a$. Hence, $bab^{-1}$ must be the other element of $H$ of order 6, namely $a^{-1} = a^5$. If $b$ has order 3 or 6 then $b^3 = e$ or $b^3 = a^3 = d$, the unique element of order 2 in $G$. In both cases $b^3ab^{-3} = a$. However, as conjugation by $b$ is a homomorphism

$$b^3ab^{-3} = b^2a^{-1}b^{-2} = bab^{-1} = a^{-1}.$$

Hence it is impossible for $b$ to have order 3 or 6 and $b$ must have order 4.

To summarize, $G$ has two elements $a, a^{-1}$ of order 6, two elements of order 3, one element, $d = a^3$ of order 2 and six elements of order 4. The first exercise below constructs such a group $G$ by the same method we used to construct the quaternion group $Q$ as a stabilizer. The quaternion group $Q$ and this group $G$ are the first two in an infinite family of so-called **dicyclic** groups. You should not attempt it without being familiar with the arguments used in the discussion of $Q$. The second exercise gives sufficient information to construct the Cayley table of any group of order 12 not covered by the three theorems above and shows that such a group must be isomorphic to $G$. These two exercises complete the classification of groups of order 12.

## EXERCISE 3

Let $p$ be the polynomial

$$\begin{aligned} p \;=\; & x_1x_2^2x_5^2x_6^3 + x_2x_3^2x_6^2x_7^3 + x_3x_4^2x_7^2x_8^3 + x_4x_1^2x_8^2x_5^3 \\ + \; & x_5x_{12}^2x_9^2x_2^3 + x_6x_9^2x_{10}^2x_3^3 + x_7x_{10}^2x_{11}^2x_4^3 + x_8x_{11}^2x_{12}^2x_1^3 \\ + \; & x_9x_8^2x_3^2x_{12}^3 + x_{10}x_5^2x_4^2x_9^3 + x_{11}x_6^2x_1^2x_{10}^3 + x_{12}x_7^2x_2^2x_{11}^3. \end{aligned}$$

(i) Show that the permutations

$$b = (1\ 2\ 3\ 4)(5\ 6\ 7\ 8)(9\ 10\ 11\ 12) \text{ and } a = (1\ 5\ 9\ 3\ 7\ 11)(12\ 8\ 4\ 10\ 6\ 2),$$

of orders 4 and 6 respectively, are in stab$(p)$. Hence, the 12 permutations $a^i, ba^i$, where $0 \leqslant i \leqslant 5$, are in stab$(p)$. Find the cycle decomposition of each of these 12 permutations. Check that six of them have order 4 and that there is a unique element of order 2.

(ii) Show that stab$(p)$ has order 12.

(iii) Check that $a^3 = b^2$ is the unique element of order 2 and that $ab = ba^5$.

### EXERCISE 4

Let $G$ be any group of order 12 not isomorphic to any of the groups $C_{12}, C_6 \times C_2, D_6, A_4$. Thus, the numbers of elements in $G$ of orders 2, 3, 4 and 6 are 1, 2, 6 and 2 respectively. Let $a$ and $b$ be elements of $G$, orders 6 and 4 respectively. (Thus, $a^3 = b^2$ is the unique element of order 2.) Show that $ab$ has order 4 and hence that $(ab)^2 = b^2$. Hence show that $ab = ba^5$.

## 17.5 Further Exercises on Chapter 17

### EXERCISE 5

Find a formula expressing an arbitrary reflection $\mathrm{ref}_{\frac{2m\pi}{p}}$ in the dihedral group $D_p$ in the form $a^i b^j$ where $a = \mathrm{rot}_{\frac{2\pi}{p}}$ and $b = \mathrm{ref}_0$.

### EXERCISE 6

Let $G$ be a group of order 15. Let $a, b$ be elements of $G$ of orders 5,3 respectively.

(i) Use Theorem 2 of Chapter 14 to show that $G$ cannot have two different subgroups of order 5. Deduce that $< a >$ must be normal in $G$.

(ii) Show that $bab^{-1} = a^j$ for some $j, 0 \leqslant j \leqslant 4$. By considering $b^3ab^{-3}$, decide which values of $j$ are possible.

(iii) Up to isomorphism, how many groups of order 15 are there?

### EXERCISE 7

Let $p_1$ be the polynomial

$$x_1x_2^2x_5^2x_6^3 + x_2x_3^2x_6^2x_7^3 + x_3x_4^2x_7^2x_8^3 + x_4x_1^2x_8^2x_5^3.$$

Show that the stabilizer of $p_1$ in $S_8$ is the cyclic subgroup generated by (1 2 3 4)(5 6 7 8).

### EXERCISE 8

Let $p_2$ be the polynomial

$$x_1x_2^2x_5^2x_6 + x_2x_3^2x_6^2x_7 + x_3x_4^2x_7^2x_8 + x_4x_1^2x_8^2x_5$$
$$+x_5x_8^2x_3^2x_2 + x_6x_5^2x_4^2x_3 + x_7x_6^2x_1^2x_4 + x_8x_7^2x_2^2x_1.$$

Show that the stabilizer of $p_2$ in $S_8$ has order 16 and that it has $Q$ as a subgroup.

## EXERCISE 9

Let $p_3$ be the polynomial

$$x_1x_2^2x_5^2x_6 + x_2x_3^2x_6^2x_7 + x_3x_4^2x_7^2x_8 + x_4x_1^2x_8^2x_5.$$

Show that the stabilizer of $p_3$ in $S_8$ has order 8. Is it isomorphic to $Q$ or to $D_4$?

## EXERCISE 10

For each of the groups $G$ of order 12, find the centre $Z(G)$. Show that if $Z(G)$ is proper and non-trivial then $G/Z(G) \cong D_3$.

# 18 • Past and Future

The aim of this chapter is to describe how group theory has developed into a central branch of mathematics. This development is still going on; if you visit the periodicals section of a university library and look at the titles of papers in the most recently published journals you will find the word 'group' in many of the titles.

The first section is a brief historical survey of how the subject grew in the nineteenth century. The other sections involve suggestions for further study either through reading new topics, for which we hope this book will have prepared you, or through projects based on the content of this book. At the end there is a bibliography listing books where you can read more.

## 18.1 History

The group theory we have covered developed principally in the nineteenth century. However, some of the ideas had previously appeared in the work of Fermat, Euler and Gauss on numbers, and of Lagrange and others on the theory of equations.

**Pierre de Fermat** (1601-1665) wrote down many theorems but rarely gave proofs. His most famous 'theorem', Fermat's Last Theorem, (see Allenby and Redfern for details) remained unproven until 1993 when Andrew Wiles announced a proof. His little theorem was first proved by **Leonhard Euler** (1707-1783) who gave several proofs, one of which effectively used the method of partition used in the proof of Lagrange's Theorem in Chapter 10. **Carl Friedrich Gauss** (1777-1855) was one of the greatest mathematicians of all time. One of his earliest interests was in roots of unity and his analysis of these included the determination of what we now call the subgroups of a finite cyclic group.

**Comte Joseph Louis de Lagrange** (1736-1813) worked in several areas of mathematics including the theory of equations. Lagrange's biggest problem in the theory of equations was the search for a formula for the roots of a quintic equation in terms of the coefficients using addition, subtraction, multiplication, division and extraction of roots. He had used orbits of polynomials in several variables to re-derive the known formulae for quadratics, cubics and quartics by a common method. In his desire to extend this method to higher degree polynomials, he experimented with the orbits of various polynomials in $n$ variables. Having observed that the number of elements in an orbit was always a factor of $n!$, he proved that this is always the case. This is the special case of the theorem which we have called the finite form of the Orbit-Stabilizer Theorem in Chapter 11 and which, in turn, was a special case of the theorem now known as Lagrange's Theorem. The reason that the general theorem bears the name of Lagrange, although Lagrange did not recognize the group concept, is that Lagrange's method of proof, by partitioning $S_n$ into the cosets $\text{send}_p(q)$ for the different polynomials $q$ in the orbit of $p$, is the same as the one used in the general case. The first person to write down and prove Lagrange's Theorem in its modern form was **Camille Jordan** (1838-1922, no relation!).

**Paolo Ruffini** (1765-1822) worked at the end of the eighteenth century and his work contains the germs of the notions of cyclic permutation, cyclic subgroup and stabilizer.

He observed different symmetry in polynomials such as $x_1x_2^2 + x_3x_4^2$ and $x_1x_2^2 + x_2x_3^2 + x_3x_4^2 + x_4x_1^2$. In our terms, the stabilizers of these, {id, (1 3), (2 4), (1 3)(2 4)} and $< (1\ 2\ 3\ 4) >$ are the two different groups of order 4.

It is only in the nineteenth century, in the work of Cauchy, Abel and Galois that the significance of closure, and with it the concepts of group and subgroup, emerged. **Baron Augustin Louis Cauchy** (1789-1857) is mainly remembered for his work on functions of a complex variable. He worked a lot on permutations and investigated several general group theoretical ideas in the context of symmetric groups. In addition to proving the theorem which bears his name, he published papers on the order of permutations and on conjugacy classes as well as on orbits and stabilizers.

In applying group theoretical ideas to the theory of equations, it was **Niels Heinrich Abel** (1802-1829) and **Evariste Galois** (1811-1832) who had the most success. Abel was interested in groups of permutations and, in particular, the alternating groups. In one of his papers on equations the main result required a certain group to be commutative and this is why abelian groups were named in his honour. He also proved that there is no formula, of the type sought by Lagrange, for the roots of a general quintic. Galois went further and found a criterion, in terms of a certain group, now named after him, for when the roots of a given polynomial could be expressed in terms of the coefficients in the desired way. Both Abel and Galois died young, Galois in a duel and Abel of tuberculosis.

The idea of an abstract group first appeared in the work of **Arthur Cayley** (1821-1895) in the 1850s. For example, he classified the groups of order 8. The quaternions had been discovered by **Sir William Rowan Hamilton** (1805-1865) in the previous decade. Infinite groups first emerged as groups of linear transformations, including rotations and reflections in two and three dimensions. Influential mathematicians in this regard included **Felix Klein** (1849-1925) and **Sophus Lie** (1842-1899) who published a joint paper on groups of transformations in 1871. Klein's 4-group is special in that it is the only abelian non-cyclic finite group of rotations in three-dimensional space.

An axiomatic definition of group, along modern lines, was given in 1882 by **Heinrich Weber** (1842-1913). His definition was for finite groups and the cancellation laws formed one of the axioms. For finite groups it is then possible to show that a neutral element and inverses must exist. Later, the definition was adapted to cover infinite groups and the existence of a neutral element and inverses became axioms. In the same journal where Weber's paper appeared (Volume 20, *Mathematische Annalen*), **Walter von Dyck** (1856-1934) defined groups of functions under composition. As associativity is automatic for composition, von Dyck did not include it as an axiom.

Weber is also responsible for the first textbook covering group theory, although earlier books had mentioned permutations. This was *Lehrbuch der Algebra* published in 1895. The first English textbook on group theory, *Theory of Groups of Finite Order* by **William Burnside** (1852-1927) appeared in 1897. This book contains many of the theorems we have covered including the Orbit-Counting Theorem.

## 18.2 Topics for Further Study

In this section we make some suggestions for further study, beginning with some individual topics in group theory.

## Frieze and wallpaper patterns

In our discussions of symmetry we have concentrated on rotations and reflections of geometrical figures. These are not the only types of symmetry that can be seen in the Euclidean plane. For example, consider the pattern

This has a **translation** $t$ which moves each diamond one place to the right. This translation is an element of the group $G$ of symmetries of the pattern. (This is not a subgroup of the orthogonal group $O_2$.) The inverse of $t$ moves each diamond one place to the left. The group $G$ also contains rotations about the centre of each diamond and about midpoints between diamonds. Moreover, there are reflections in a horizontal line and in infinitely many vertical lines. Notice that the translation $t$ does not commute with rotations : try computing $tr$ and $rt$ where $r$ is rotation through $\pi$ about the centre of one of the diamonds. Another type of symmetry, a **glide** is shown by the following pattern.

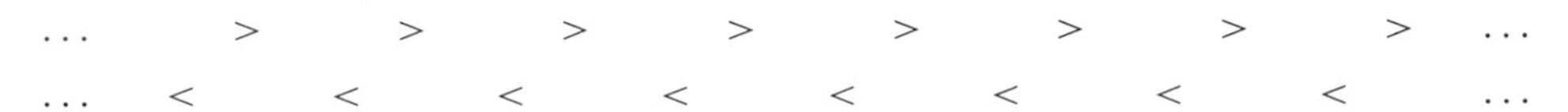

The glide symmetry $g$ is the composition of a reflection (in a vertical line) and a vertical translation, which are not themselves symmetries. These patterns are examples of **frieze patterns**. There are essentially seven different frieze patterns classified by their groups of symmetries. The book by Martin has a chapter on these (see Bibliography, page 204). If we allow translations in a second direction we obtain the so-called **wallpaper patterns** of which there are essentially 17 different groups of symmetry. Further details of these can be found in the books by Allenby (1991), Armstrong, Artin and Martin.

## The Sylow Theorems

In our classification of groups of small order, we made use of Cauchy's Theorem which stated that if $p$ is a prime number dividing the order of a finite group $G$ then $G$ must have an element, and hence a subgroup, of order $p$. Stronger theorems of this kind were obtained by **Ludwig Sylow** (1832-1918). He proved that, if $p^k$ is the largest power of $p$ dividing $|G|$, then $G$ must have a subgroup of order $p^k$. Any such subgroup is called a Sylow $p$-subgroup. Sylow also showed that any two Sylow $p$-subgroups are conjugate and gave information on the number of Sylow $p$-subgroups that can occur. Sometimes one can quickly conclude that there is only one and that it is normal. This helps greatly in the classification of groups of a given order. These theorems are now accessible to you, particularly if you read a proof using group actions as, for example, in the books by Armstrong and Artin.

## Generators and relations

Whereas every element of a cyclic group can be written as a power of a single element, every element of the dihedral group $D_n$ can be written as a product of powers of two elements, the rotation $a = \text{rot}_{\frac{2\pi}{n}}$ and the reflection $b = \text{ref}_0$ . In such circumstances, $\{a, b\}$ is said to be a set of **generators** for the group. Multiplication of elements of the form $b^i a^j$ can be performed using the equations $a^n = e$, $b^2 = e$ and $ab = ba^{n-1}$. These are called **relations**. In several places in Chapter 17 we were, in effect, using generators and relations.

Sometimes groups are specified by generators and relations. We can then calculate with the elements and even program a computer to do so. There is a drawback in that if you write down some generators and relations at random it may not be clear whether they determine a finite group. The book by Artin discusses the theory of generators and relations, including **free groups** which are groups with no relations.

## Rings and fields

We have seen several sets, notably $\mathbb{R}$, $\mathbb{Z}$, $\mathbb{C}$, $\mathbb{Q}$ and, for each integer $n > 1$, $\mathbb{Z}_n$, where there are two binary operations, addition and multiplication. These are examples of **rings** which, like groups, are defined by a list of axioms. The axioms include the distributive law $a(b + c) = ab + ac$ on combining the two operations.

Each of the rings $R$ listed above is an abelian group under addition with neutral element 0. In each case we also obtained a group under multiplication by taking those elements of $R$ with multiplicative inverses in $R$. This is the **group of units** $U(R)$ of $R$. Sometimes this consists of all the non-zero elements of $R$ and in this case $R$ is called a **division ring**. In the examples listed, multiplication is always commutative but this need not be so. Examples of non-commutative rings include $2 \times 2$ real matrices and Hamilton's quaternions described in Chapter 17. The latter is a division ring but the former is not as there are non-zero non-invertible $2 \times 2$ matrices. The group of units for $2 \times 2$ real matrices is the general linear group $GL_2(\mathbb{R})$. A commutative division ring is called a **field**. Examples include $\mathbb{R}$, $\mathbb{C}$, $\mathbb{Q}$ and, for each prime $p$, $\mathbb{Z}_p$.

The group of units is a nice example of interplay between groups and rings. This features in the classification of finite fields, one of which exists for each prime power $p^n$. A key step in this classification involves showing that the group of units of a finite field is cyclic.

## Finite abelian groups

In our classification of groups up to order 12 in Chapter 17, all the abelian groups are cyclic or direct products of cyclic groups. This is true of all finite abelian groups. You can read about this in group theoretical terms in the book by Allenby (1991). Alternatively, this classification is often presented as the specialization to $\mathbb{Z}$ of a more general theory involving certain commutative rings. This approach is taken in the book by Artin.

## Galois theory

Whenever we have one field $F$ inside another field $K$, for example $\mathbb{R} \subset \mathbb{C}$, there is a group called the **Galois group**, $\text{Gal}(K/F)$, of $K$ over $F$. This consists of all bijective functions $\theta : K \to K$ such that $\theta(c) = c$ for all $c \in F$ and, for all $a, b \in K$, $\theta(a + b) = \theta(a) + \theta(b)$ and $\theta(ab) = \theta(a)\theta(b)$.

If $f(x)$ is a polynomial in $x$ with coefficients in a field $K$ then the Galois group of $f(x)$ over $F$ is $\text{Gal}(K/F)$ where $K$ is in some sense the smallest field containing both $F$ and the roots of $f(x)$. The Galois group then acts on the roots of $f(x)$ and is isomorphic to a subgroup of $S_n$ where $n$ is the number of distinct roots. When $F = \mathbb{R}$ and $f(x) = x^2 + 1$, the roots are $\pm 1$, the field $K$ is $\mathbb{C}$ and the Galois group is isomorphic to $S_2$. The elements are the two functions from $\mathbb{C}$ to itself with $a + bi \mapsto a + bi$ and $a + bi \mapsto a - bi$.

The criterion we referred to in the previous section is called solubility. Loosely speaking, the soluble groups are the groups which can be built from abelian groups using normal subgroups and factor groups. The first non-abelian soluble groups are groups $G$, such as $S_3$, with a normal subgroup $N$ such that both $N$ and $G/N$ are abelian. The group $S_4$ is also soluble but $S_5$ is not, basically because $A_5$ is simple. This explains the difference between quartic and quintic equations.

### Representation theory

If you have enjoyed this introduction to groups and are familiar with linear algebra, we recommend that you look for the chance to take a course on representation theory. It is this branch of group theory that is most readily applicable outside mathematics. For example, it has been used successfully to predict the existence of new elementary particles in physics. (You can read more about this application in the book by Kostrikin and Shafarevich.)

A **representation** of a group $G$ is an action of $G$ on a vector space $V$ in such a way that, for each $g \in G$, the function $v \mapsto g * v$ is a linear transformation. This can also be formulated in terms of homomorphisms from $G$ to general linear groups. Any group action can be used to obtain a representation using the idea of a permutation matrix, this being a matrix where each entry is 1 or 0 and there is exactly one 1 in each row and column.

Taking the trace of each of the linear transformations $v \mapsto g * v$ gives the **character** of a representation. For finite groups, the characters can be displayed in a **character table** which carries much information about the group, including the conjugacy classes, and can be helpful in classification problems. For example, Burnside used this approach to show that a finite group of order $p^\alpha q^\beta$, where $p$ and $q$ are distinct primes, cannot be simple.

### Continuous groups

We have seen that the matrices $\begin{pmatrix} \cos\theta & -\sin\theta \\ \sin\theta & \cos\theta \end{pmatrix}$, where $\theta$ is a real parameter, form a group, the special orthogonal group $SO_2$. This is an example of a **one-parameter** group. Another example is the group of matrices of the form $\begin{pmatrix} 1 & a \\ 0 & 1 \end{pmatrix}$, where $a$ is a real parameter. In these groups, and in other subgroups of $GL_n(\mathbb{R})$ where more parameters are needed, the parameters can vary continuously. Groups such as these play important roles in analysis, geometry and topology. The main groups used in differential geometry are named after Lie who first recognized this unifying role for groups. Subgroups of general linear groups are called **linear groups** and you can read about them in Artin's book.

## 18.3 Projects

The projects listed below cover various aspects of group theory. Although we have sometimes asked specific questions, we hope that you will come up with questions of your own. For some of the projects, in particular those on symmetries of frieze patterns and three-dimensional objects, it will be helpful to refer to other texts, possibly after a preliminary investigation. The projects are not intended only for individual study. There is a lot to be gained from discussion of mathematical problems with others.

1. At the end of Chapters 1 and 11, we suggested that you investigate orbits for the action of $D_4$ on various shapes in a square grid. Try the same idea for triangular and other patterns

in two dimensions or for the action of the rotation group of the cube on a cube divided into small cubes.

**2.** Investigate the rotation groups of the icosahedron and dodecahedron.

(i) Are they isomorphic and, if so, why?

(ii) The action of the rotation group of the cube on the diagonals gives rise to an isomorphism between the rotation group and the alternating group $A_4$. Can you find a similar situation for these groups?

(iii) Investigate conjugacy classes for these groups.

**3.** How many different frieze and wallpaper patterns can you find?

**4.** In Chapter 13 we looked at conjugacy in the orthogonal group and found, for example, a formula for the conjugate of a reflection by a rotation as another reflection. Investigate conjugacy with regard to translations, glides, rotations and reflections in groups of symmetries of frieze patterns.

**5.** If $p$ is a prime number then $\mathbb{Z}_p$ is a field and, as for $\mathbb{R}$, we can form the general linear group $GL_2(\mathbb{Z}_p)$ and the special linear group $SL_2(\mathbb{Z}_p)$. Investigate these groups, at least for small primes.

(i) Find the orders of these groups and of some elements.

(ii) Find the centres of these groups.

(iii) The factor group $SL_2(\mathbb{Z}_p)/Z(SL_2(\mathbb{Z}_p))$, which is written $PSL_2(\mathbb{Z}_p)$ ($PSL$ stands for projective special linear), is important. Find its order. Can you find some projective special linear groups which are isomorphic to familiar groups? Can you find one of order 168?

**6.** In Chapter 17, we constructed the first two dicyclic groups which have orders 8 and 12. Can you spot a sufficient pattern in their construction to construct the next one as a subgroup of a symmetric group, as the stabilizer of a polynomial or in terms of generators and relations?

**7.** In the last exercise of Chapter 10 we asked you to investigate the subgroups of $S_4$. Take this further. Which subgroups are conjugate to each other? Which of them can be written as stabilizers of polynomials in four variables? Can you prove that you have found them all?

**8.** Write computer programs to

(i) compute products and conjugates in orthogonal and dihedral groups;

(ii) draw orbits for actions of groups of matrices on $\mathbb{R}^2$;

(iii) find the cycle decomposition of a permutation.

**9.** Find out more about some of the mathematicians mentioned in Section 18.1 and their work.

**10.** Example (v) of Section 7.2 shows the action of $(\mathbb{R}, +)$ on $\mathbb{R}^2$ given by $t * (x, y) = (x + ty, y)$. Actions of $(\mathbb{R}, +)$ on $\mathbb{R}^2$ are called **flows** and can give rise to some attractive orbits. Investigate the following examples of flows. Check that they are group actions. Find orbits for different values of the constants $a$ and $b$ including both positive and negative values. This will be best done using a computer or graphical calculator that can plot curves given in parametric form. Can you find any other examples of flows?

(i) $t * (x, y) = (xe^{at}, ye^{bt})$;

(ii) $t * (x, y) = ((x + aty)e^{bt}, ye^{bt})$;

(iii) $t * (x, y) = (e^{at}(x \cos(bt) + y \sin(bt)), e^{at}(y \cos(bt) - x \sin(bt)))$.

**11.** The group stab$(M)$ in (iii) of the last exercise in Chapter 13 is 'similar' to the orthogonal group $O_2$. Compare these groups, describing the features they have in common. Are they isomorphic? If you know about the hyperbolic functions sinh and cosh, what is their relevance here?

**12.** An isomorphism from a group $G$ to itself is called an **automorphism** of $G$. Investigate automorphisms of particular groups, beginning with the cyclic group $C_3$. Do the automorphisms of a group form another group? If so, investigate its order for some particular groups. Can you suggest some constructions of automorphisms for all groups or for all abelian groups?

**13.** Investigate groups of order 21. What is different to the situation in Exercise 6 of Chapter 17 for groups of order 15? How can Exercise 10 of Chapter 13 be used to construct a non-cyclic group of order 21? Can you find a polynomial in seven variables with a non-cyclic stabilizer of order 21? Investigate groups of other orders which can be constructed by adapting Exercise 10 of Chapter 13. What order is the group that results from replacing $\alpha^2$ by $\alpha^3$?

# Solutions

## Chapter 1

**1.** $s_3 r_1 = s_2$; $r_1 r_2 = r_3$; $s_1 s_2 = r_3$.

**2.** $\text{ref}_\alpha \text{rot}_\beta(r\cos\theta, r\sin\theta) = \text{ref}_\alpha(r\cos(\theta+\beta), r\sin(\theta+\beta))$
$= (r\cos(\alpha-(\theta+\beta)), r\sin(\alpha-(\theta+\beta))) = \text{ref}_{\alpha-\beta}(r\cos\theta, r\sin\theta)$.

**3.** $r_1 s_1 = s_2$, $(r_1 s_1)s_3 = s_2 s_3 = r_3$, $s_1 s_3 = r_2$, $r_1(s_1 s_3) = r_1 r_2 = r_3$. Yes.

**4.** $\text{stab}(2) = \{e, s_4\}$, $\text{send}_2(1) = \{s_3, r_3\}$, $\text{send}_2(3) = \{s_1, r_1\}$, $\text{send}_2(4) = \{s_2, r_2\}$.

**5.** (i) (a) $\text{stab}(2) = \{e, s_3\}$ (b) $\text{stab}(5) = D_4$;

(ii) (a) $\text{orb}(2) = \{2, 4, 6, 8\}$ (b) $\text{orb}(5) = \{5\}$;

(iii) 8 both times;

(iv) No. of elements in $D_4$ = no. of elements in stabilizer × no. of elements in orbit. Yes.

**6.** (i) $\{\text{rot}_0, \text{ref}_\pi\}$; (ii) $\{\text{rot}_{\frac{\pi}{2}}, \text{ref}_{\frac{3\pi}{2}}\}$; (iii) 2,$\{\text{rot}_{\phi-\theta}, \text{ref}_{\phi+\theta}\}$.

## Chapter 2

**1.** $gf = \begin{pmatrix} 1 & 2 & 3 & 4 \\ 3 & 4 & 2 & 1 \end{pmatrix}$. No.

**2.** $\begin{pmatrix} 1 & 2 & 3 & 4 \\ 3 & 4 & 2 & 1 \end{pmatrix}$, $\begin{pmatrix} 1 & 2 & 3 & 4 \\ 3 & 4 & 1 & 2 \end{pmatrix}$.

**3.** (i) (a) 4,6 (b) 1,24 (c) 6,4.

(ii)

| | id | $f$ | $g$ | $h$ |
|---|---|---|---|---|
| id | id | $f$ | $g$ | $h$ |
| $f$ | $f$ | id | $h$ | $g$ |
| $g$ | $g$ | $h$ | id | $f$ |
| $h$ | $h$ | $g$ | $f$ | id |

**4.** (i) $f^{-1}(x) = x+3$; $g^{-1}(x) = \frac{x}{2}$.

(ii) $fg(x) = 2x-3$; $gf(x) = 2x-6$; $f^{-1}g^{-1}(x) = \frac{x}{2}+3$; $g^{-1}f^{-1}(x) = \frac{x+3}{2}$.

(iii) $(fg)(f^{-1}g^{-1})(x) = x+3$; $(fg)(g^{-1}f^{-1})(x) = x$; $(f^{-1}g^{-1})(fg)(x) = x+\frac{3}{2}$; $(g^{-1}f^{-1})(fg)(x) = x$. $(fg)^{-1} = g^{-1}f^{-1}$ and $(gf)^{-1} = f^{-1}g^{-1}$.

**5.**

| Domain $A$ | Codomain $B$ | $f(a)$ | surj? yes/no | inj? yes/no | bij? yes/no | $b$ with, for all $a$, $f(a) \neq b$ | $a_1, a_2$ with $a_1 \neq a_2$ $f(a_1) = f(a_2)$ |
|---|---|---|---|---|---|---|---|
| $\{x \in \mathbb{R} : x \geqslant 0\}$ | $\mathbb{R}$ | $a^2$ | no | yes | no | $-1$ | none |
| $\mathbb{R}$ | $\{x \in \mathbb{R} : x \geqslant 0\}$ | $a^2$ | yes | no | no | none | $-1, 1$ |
| $\mathbb{R}$ | $\mathbb{R}$ | $a^3$ | yes | yes | yes | none | none |
| $\mathbb{R}$ | $\mathbb{R}$ | $a^4$ | no | no | no | $-1$ | $-1, 1$ |
| $\mathbb{R}\backslash\{0\}$ | $\mathbb{R}\backslash\{0\}$ | $\frac{1}{a}$ | yes | yes | yes | none | none |
| $\mathbb{R}$ | $O_2$ | $\text{rot}_a$ | no | no | no | $\text{ref}_0$ | $0, 2\pi$ |
| $\mathbb{R}$ | $O_2$ | $\text{ref}_a$ | no | no | no | $\text{rot}_\pi$ | $0, 2\pi$ |

**6.** $fg = \begin{pmatrix} 1 & 2 & 3 & 4 & 5 \\ 4 & 5 & 2 & 1 & 3 \end{pmatrix}$; $gf = \begin{pmatrix} 1 & 2 & 3 & 4 & 5 \\ 2 & 5 & 4 & 3 & 1 \end{pmatrix}$;

$f^{-1} = \begin{pmatrix} 1 & 2 & 3 & 4 & 5 \\ 3 & 5 & 2 & 4 & 1 \end{pmatrix}$; $f^{-1}g = \begin{pmatrix} 1 & 2 & 3 & 4 & 5 \\ 4 & 3 & 1 & 2 & 5 \end{pmatrix}$.

**7.** $\begin{pmatrix} 1 & 2 & 3 \\ 1 & 2 & 3 \end{pmatrix}$; $\begin{pmatrix} 1 & 2 & 3 \\ 2 & 3 & 1 \end{pmatrix}$; $\begin{pmatrix} 1 & 2 & 3 \\ 3 & 1 & 2 \end{pmatrix}$; $\begin{pmatrix} 1 & 2 & 3 \\ 2 & 1 & 3 \end{pmatrix}$; $\begin{pmatrix} 1 & 2 & 3 \\ 3 & 2 & 1 \end{pmatrix}$; $\begin{pmatrix} 1 & 2 & 3 \\ 1 & 3 & 2 \end{pmatrix}$.

**8.** $4! = 24$

**9.** (i) (a) $\begin{pmatrix} 1 & 2 & 3 \\ 2 & 3 & 1 \end{pmatrix}$ for both. (b) $\begin{pmatrix} 1 & 2 & 3 & 4 & 5 \\ 2 & 3 & 4 & 5 & 1 \end{pmatrix}$; $\begin{pmatrix} 1 & 2 & 3 & 4 & 5 \\ 5 & 1 & 2 & 3 & 4 \end{pmatrix}$.

(ii) $f^2 = (1\ 3\ 2)$, $f^3 = \text{id}$, $f^k = \text{id}$.

(iii) $fg = \text{id}$, $gf = \text{id}$, $(a_1\ a_2 \ldots a_k)^{-1} = (a_k \ldots a_2\ a_1)$.

(iv) $fg = \begin{pmatrix} 1 & 2 & 3 & 4 & 5 & 6 \\ 2 & 4 & 5 & 1 & 6 & 3 \end{pmatrix} = gf$.

(v) $hg = \begin{pmatrix} 1 & 2 & 3 & 4 & 5 & 6 \\ 2 & 3 & 5 & 4 & 6 & 1 \end{pmatrix}$, $gh = \begin{pmatrix} 1 & 2 & 3 & 4 & 5 & 6 \\ 2 & 5 & 1 & 4 & 6 & 3 \end{pmatrix}$. No.

**10.** (i) $f = (1\ 2\ 9\ 4)(3\ 7\ 8)(5)(6\ 10)$.

(ii) (a) $(1\ 2)(3)(4\ 5)$; (b) $(1\ 2\ 3\ 4\ 5)$; (c) $(1\ 2\ 3\ 4\ 5)$; (d) $(1\ 4)$.

**11.** $(1\ 2\ 9\ 4)(3\ 7\ 8)(5)(6\ 10) = (1\ 2)(2\ 9)(9\ 4)(3\ 7)(7\ 8)(6\ 10)$
$= (1\ 4)(1\ 9)(1\ 2)(3\ 8)(3\ 7)(6\ 10)$.

**12.** (i) $(2\ 4) = (1\ 2)(1\ 4)(1\ 2)$, $(i\ j) = (1\ i)(1\ j)(1\ i)$.

(ii) $(1\ 4)(1\ 9)(1\ 2)(1\ 8)(1\ 3)(1\ 8)(1\ 7)(1\ 3)(1\ 7)(1\ 10)(1\ 6)(1\ 10)$.

## Chapter 3

**1.** (i) (a) $AB = \begin{pmatrix} 0 & 3 \\ -5 & 2 \end{pmatrix}$; $BA = \begin{pmatrix} 1 & 7 \\ -2 & 1 \end{pmatrix}$.

(b) $AB = \begin{pmatrix} 0 & 0 \\ 0 & 0 \end{pmatrix}$; $BA = \begin{pmatrix} 0 & 0 \\ 0 & 0 \end{pmatrix}$.

(ii) $BX = \begin{pmatrix} x + y \\ y \end{pmatrix}$; $A(BX) = \begin{pmatrix} 2(x + y) \\ 2y \end{pmatrix}$; $AB = \begin{pmatrix} 2 & 2 \\ 0 & 2 \end{pmatrix}$;

$(AB)X = \begin{pmatrix} 2(x + y) \\ 2y \end{pmatrix}$.

(iii) $AB = \begin{pmatrix} ae+bg & af+bh \\ ce+dg & cf+dh \end{pmatrix}$;

$(AB)C = \begin{pmatrix} aep+bgp+afr+bhr & aeq+bgq+afs+bhs \\ cep+dgp+cfr+dhr & ceq+dgq+cfs+dhs \end{pmatrix}$;

$BC = \begin{pmatrix} ep+fr & eq+fs \\ gp+hr & gq+hs \end{pmatrix}$; $A(BC) = (AB)C$.

(iv) Routine.

**2.** (i) $\begin{pmatrix} 2 & -1 \\ -1 & 1 \end{pmatrix}$; (ii) $\begin{pmatrix} 1 & -\frac{1}{2} \\ 2 & -\frac{1}{2} \end{pmatrix}$; (iii) no inverse.

**3.** $f_A \circ f_B\left(\begin{pmatrix} x \\ y \end{pmatrix}\right) = f_A\left(\begin{pmatrix} x+ay \\ y \end{pmatrix}\right) = \begin{pmatrix} 2(x+ay) \\ 2y \end{pmatrix}$.

$f_{AB}\left(\begin{pmatrix} x \\ y \end{pmatrix}\right) = \begin{pmatrix} 2 & 2a \\ 0 & 2 \end{pmatrix}\begin{pmatrix} x \\ y \end{pmatrix} = \begin{pmatrix} 2(x+ay) \\ 2y \end{pmatrix}$.

**4.** $f_A\left(\begin{pmatrix} x \\ y \end{pmatrix}\right) = \begin{pmatrix} x \\ 2x+y \end{pmatrix}$. This is a shear parallel to the $x$-axis.

$f_A\left(\begin{pmatrix} x \\ y \end{pmatrix}\right) = \begin{pmatrix} -x \\ -y \end{pmatrix}$. This is rotation through $\pi$. $f_A\left(\begin{pmatrix} x \\ y \end{pmatrix}\right) = \begin{pmatrix} y \\ x \end{pmatrix}$. This is reflection in $y = x$.

**5.** (i) $A^T = \begin{pmatrix} 1 & 2 & 3 \\ 2 & -1 & 7 \\ 6 & 1 & 2 \end{pmatrix}$. (ii) $(A^T)^T = A$.

**6.** (i) Yes. $A_\alpha A_\beta = A_{\alpha+\beta}$; $B_\alpha B_\beta = A_{\alpha-\beta}$; $A_\alpha B_\beta = B_{\alpha+\beta}$; $B_\alpha A_\beta = B_{\alpha-\beta}$.
(ii) Yes. $A_\alpha^{-1} = A_{-\alpha}$; $B_\alpha^{-1} = B_\alpha$.

## Chapter 4

**1.** (i) Addition is a binary operation on $A$. Neither multiplication nor subtraction is since $-1 \times -2 = 2 \notin A$ and $-1 - (-2) = 1 \notin A$.

(ii) If $ab + a + b = -1$ then $a(b+1) + (b+1) = 0$ so $(a+1)(b+1) = 0$ and hence $a = -1$ or $b = -1$.

**2.** $g^2 = g \Rightarrow g^2g^{-1} = gg^{-1} \Rightarrow g = e$.

**3.** (i) $x = \text{ref}_{\frac{\pi}{3}}$; $y = \text{ref}_{\frac{2\pi}{3}}$. (ii) $x = s_4$; $y = s_2$. (iii) $x = (1\ 3\ 4\ 2)$; $y = (1\ 4\ 2\ 3)$.

**4.** $g = s_1,\ h = s_2,\ k = s_4,\ gh = r_3 = kg$.

**5.** $ab = (ab)^{-1} = b^{-1}a^{-1} = ba$.

**6.** $(g_1, h_1)(g_2, h_2) = (g_1g_2, h_1h_2) = (g_2g_1, h_2h_1) = (g_2, h_2)(g_1, h_1)$.

## Chapter 5

**1.** (i) Yes. (ii) No since $(2\ 3)(3\ 4) = (2\ 3\ 4) \notin H$. (iii) Yes. (iv) No, since $\frac{\pi}{2} + \frac{3\pi}{2} = 2\pi \notin H$.

**2.** $\begin{pmatrix} 1 & a \\ 0 & 1 \end{pmatrix}\begin{pmatrix} 1 & b \\ 0 & 1 \end{pmatrix} = \begin{pmatrix} 1 & a+b \\ 0 & 1 \end{pmatrix}$; $\begin{pmatrix} 1 & a \\ 0 & 1 \end{pmatrix}^{-1} = \begin{pmatrix} 1 & -a \\ 0 & 1 \end{pmatrix}$. Yes it is abelian.

**3.** (i) F,G,J,P,R; (ii) A,B,C,D,E,K,L,M,Q,T,U,V,W; (iii) N,S,Z; (iv) H,I; (v) Y (vi) X: (vii) O.

## Chapter 6

**1.** (i) $\text{rot}_{\frac{\pi}{3}}$, $\text{rot}_{\frac{5\pi}{3}}$ (ii) 1, $-1$.

**2.** (i) $< g >= \{(1\,2\,3\,4), (1\,3)(2\,4), (1\,4\,3\,2), \text{id}\}$;

(ii) $< g >= \left\{\begin{pmatrix} 0 & 1 \\ 1 & 0 \end{pmatrix}, \begin{pmatrix} 1 & 0 \\ 0 & 1 \end{pmatrix}\right\}$;

(iii) $< g >= \left\{\begin{pmatrix} 1 & n \\ 0 & 1 \end{pmatrix} : n \in \mathbb{Z}\right\}$.

**3.** 6,3,2,2.

**4.** 1, $-1$.

**5.** When $n$ is odd, $n$. When $n$ is even, $n + 1$.

**6.** (i) 6; (ii) 2; (iii) 4.

**7.** (i) $(1\,9)(2\,8)(3\,7)(4\,5\,6)$ has order 6.

(ii) $(1\,2\,3\,7)(4\,9)(5)(6\,8)$ has order 4.

(iii) $(1\,2\,3\,4\,5)(6\,7\,8\,9)$ has order 20.

**8.** (i) $ab = (2\,3)$ has order $2 \neq 6$.

(ii) $ab = \text{ref}_{\frac{-2\pi}{7}}$ has order $2 \neq 14$.

(iii) $ab = (1\,5\,3)(2\,6\,4)$ has order $3 \neq 6$. $ba = (1\,5\,3)(2\,6\,4) = ab$ and $a^3 = (1\,4)(2\,5)(3\,6) = b$.

**9.** 6,4,12,4.

**10.** (i) $g^3$ has order $\frac{30}{3} = 10$ and so $< g^3 >= G$.

(ii) 20.

**11.** (i) If $g^7 \in H$ then $g = (g^2)^{-3}g^7 \in H$.

(ii) If $g^{-3} \in H$ then $g^3 = (g^{-3})^{-1} \in H$ and so $g = (g^2)^{-1}g^3 \in H$.

**12.** 8.

## Chapter 7

**1.** (i) 4, (ii) $(0, \sqrt{2})$, (iii) 2, (iv) $\begin{pmatrix} -17 \\ -10 \end{pmatrix}$.

**2.** Three orbits (corners, middle 4, mid-sides).

**3.** $\{1, 2\}, \{3, 4, 5, 7\}, \{6\}$.

**4.** $\text{stab}(1) = \{e, s_3\}$, $\text{stab}(2) = D_3$, $\text{stab}(3) = \{e, s_1\}$, $\text{stab}(4) = \{e, s_2\}$.

**5.** $f_e = \begin{pmatrix} 1 & 2 & 3 & 4 \\ 1 & 2 & 3 & 4 \end{pmatrix}$, $f_{r_1} = \begin{pmatrix} 1 & 2 & 3 & 4 \\ 4 & 2 & 1 & 3 \end{pmatrix}$, $f_{r_2} = \begin{pmatrix} 1 & 2 & 3 & 4 \\ 3 & 2 & 4 & 1 \end{pmatrix}$,
$f_{s_1} = \begin{pmatrix} 1 & 2 & 3 & 4 \\ 4 & 2 & 3 & 1 \end{pmatrix}$, $f_{s_2} = \begin{pmatrix} 1 & 2 & 3 & 4 \\ 3 & 2 & 1 & 4 \end{pmatrix}$, $f_{s_3} = \begin{pmatrix} 1 & 2 & 3 & 4 \\ 1 & 2 & 4 & 3 \end{pmatrix}$.

**6.** $f = (1\ 5\ 7\ 4)(3\ 9\ 8\ 6)$ is even with $\text{sgn}\, f = 1$ and $f \in A_9$.

**7.** $\text{sgn}(f^2) = (\text{sgn}\, f)^2 = (\pm 1)^2 = 1$ for all permutations $f$. Since $(1\ 2\ 3\ 4\ 5\ 6)$ has sign $-1$ it cannot be the square of any permutation in $S_6$.

## Chapter 8

**1.** $\{e, s_3\}, \{r_1, s_4\}, \{r_2, s_1\}, \{r_3, s_2\}$.

**2.** $\{\ldots, -8, -4, 0, 4, 8, \ldots\}$, $\{\ldots, -7, -3, 1, 5, 9, \ldots\}$
$\{\ldots, -6, -2, 2, 6, 10, \ldots\}$ $\{\ldots, -5, -1, 3, 7, 11, \ldots\}$.

**3.** Reflexivity: $aRa$ since $a - a = 0 \in \mathbb{Z}$. Symmetry: if $aRb$ then $a - b \in \mathbb{Z}$. Thus, $b - a = -(a - b) \in \mathbb{Z}$ and $bRa$. Transitivity: if $aRb$ and $bRc$ then $a - b \in \mathbb{Z}$ and $b - c \in \mathbb{Z}$ and $a - c = (a - b) + (b - c) \in \mathbb{Z}$ and $aRc$.

**4.** $\{n + \frac{3}{4} : n \in \mathbb{Z}\}$ which is the orbit of $\frac{3}{4}$ under the given action.

**5.** The lines $x = c$ for $c \neq 0$ and the points $(0, b)$.

**6.** $\text{send}_4(4) = \{e, s_1\}$, $\text{send}_4(2) = \{r_3, s_4\}$, $\text{send}_4(6) = \{r_2, s_3\}$, $\text{send}_4(8) = \{r_1, s_2\}$.

**7.** (i) $a^{-1}b * x = a^{-1} * y = a^{-1} * (a * x) = a^{-1}a * x = e * x = x$. Thus, $a^{-1}b = h$ for some $h \in H$ and so $b = ah$.

(ii) $ah * x = a * (h * x) = a * x = y$.

**8.** $U(\mathbb{Z}_3) = \{\overline{1}, \overline{2}\}$, $U(\mathbb{Z}_4) = \{\overline{1}, \overline{3}\}$, $U(\mathbb{Z}_8) = \{\overline{1}, \overline{3}, \overline{5}, \overline{7}\}$.

## Chapter 9

**1.** $\det\begin{pmatrix} 1 & 0 \\ 0 & 1 \end{pmatrix} = \det\begin{pmatrix} -1 & 0 \\ 0 & -1 \end{pmatrix}$ so $f$ is not injective. Since $r = \det\begin{pmatrix} r & 0 \\ 0 & 1 \end{pmatrix}$ for all $r \in \mathbb{R}\backslash\{0\}$, $f$ is surjective.

**2.** No permutation has sgn 2. $f(1\ 2\ 3) = 1 = f(\text{id})$ so $f$ is not injective. Yes it is injective when $n = 2$.

**3.** $f(x_1 + x_2) = e^{x_1 + x_2} = e^{x_1}e^{x_2} = f(x_1)f(x_2)$. Hence, $f$ is a homomorphism. It is bijective since it has an inverse, $f^{-1}(x) = \ln x$. Now $\ln(x_1x_2) = \ln(x_1) + \ln(x_2)$ and so $f^{-1}$ is an isomorphism.

**4.** Since $f$ and $g$ are bijections $gf$ is a bijection. $gf(g_1g_2) = g(f(g_1)f(g_2))$ since $f$ is an isomorphism, and $g(f(g_1)f(g_2)) = gf(g_1)gf(g_2)$ since $g$ is an isomorphism. Thus, $gf(g_1g_2) = gf(g_1)gf(g_2)$ and so $gf$ is an isomorphism and $G \cong K$.

**5.** $\mathrm{rot}_{\frac{\pi}{2}}$ has order 4 and Klein's 4-group has no such element. Apply Theorem 3.

**6.** $g: e \mapsto \mathrm{id}, r_1 \mapsto (1\,2\,3\,4), r_2 \mapsto (1\,3)(2\,4), r_3 \mapsto (1\,4\,3\,2), s_1 \mapsto (2\,4),$
$s_2 \mapsto (1\,2)(3\,4), s_3 \mapsto (1\,3), s_4 \mapsto (2\,3)(1\,4)$. $g \neq f$ but $g(D_4) = f(D_4)$.

**7.** $f_e = \mathrm{id}$, $f_{r_1} = (1\,2)$, $f_{r_2} = \mathrm{id}$, $f_{r_3} = (1\,2)$, $f_{s_1} = (1\,2)$, $f_{s_2} = \mathrm{id}$, $f_{s_3} = (1\,2)$, $f_{s_4} = \mathrm{id}$.
No it is not an injection.

**8.** $f(e) = \mathrm{id}$, $f(r_1) = (1\,2\,3\,4)(5\,6\,7\,8)$, $f(r_2) = (1\,3)(2\,4)(5\,7)(6\,8)$,
$f(r_3) = (1\,4\,3\,2)(5\,8\,7\,6)$, $f(s_1) = (1\,5)(2\,8)(3\,7)(4\,6)$, $f(s_2) = (1\,6)(2\,5)(3\,8)(4\,7)$,
$f(s_3) = (1\,7)(2\,6)(3\,5)(4\,8)$, $f(s_4) = (1\,8)(2\,7)(3\,6)(4\,5)$.

**9.** Let $f : G \to H$ be an isomorphism and let $g$ be a generator of $G$. Let $h = f(g)$. Given $k \in H$ then, for some $m \in \mathbb{Z}$, $k = f(g^m) = h^m$. Thus, $h$ generates $H$ and $H$ is cyclic.

**10.** (i) By closure $g^2 = g$ or $g^2 = e$. If $g^2 = g$ then $g = e$ so $g^2 \neq g$ and $g$ has order 2. Thus, $g$ generates $G$ and so $G$ is cyclic.

(ii) By cancellation, if $gh = g = ge$ then $h = e$ and if $gh = h$ then $g = e$. Thus, $gh = e$. If $g^2 = e = gh$ then $g = h$ and if $g^2 = g = ge$ then $g = e$. Thus $g^2 = h$ and $G = \{e, g, g^2\}$ is cyclic.

## Chapter 10

**1.** $r_1H = \{r_1, s_2\} = s_2H$, $r_2H = \{r_2, s_3\} = s_3H$, $eH = s_1H = H$.

**2.** $\mathrm{send}_L(L') = gH = \{\mathrm{rot}_{\frac{\pi}{4}}, \mathrm{rot}_{\frac{5\pi}{4}}, \mathrm{ref}_{\frac{\pi}{4}}, \mathrm{ref}_{\frac{5\pi}{4}}\}$.

**3.** $\mathrm{stab}(1) = \{e, s_1\}$, $\mathrm{send}_1(2) = \{r_3, s_4\} = r_3\,\mathrm{stab}(1)$.

**4.** $H$ has two distinct left cosets, $eH = H$ and $s_1H = \{s_1, s_2, s_3, s_4\}$ with four elements in each.

**5.** (i) 2.

(ii) If $ab = a = ae$ then $b = e$, if $ab = b = eb$ then $a = e$ and if $ab = e = aa$ then $a = b$, by cancellation. Hence, $ab = c$. Similarly, $ba = c$.

|   | $e$ | $a$ | $b$ | $c$ |
|---|---|---|---|---|
| $e$ | $e$ | $a$ | $b$ | $c$ |
| $a$ | $a$ | $e$ | $c$ | $b$ |
| $b$ | $b$ | $c$ | $e$ | $a$ |
| $c$ | $c$ | $b$ | $a$ | $e$ |

**6.** Any proper subgroup $H$ of $G$ has order 1,3 or 7. Apart from 1, which gives $< e >$, these are prime so $H$ must be cyclic. The number of elements of order 7 must be a multiple of 6 and so is at most 18. The remaining elements must each have order 3 or 21. If $g$ has order 21 then $g^7$ has order 3. Either way there is an element of order 3.

**7.** $297 \equiv 2 \bmod 5$, $297 \equiv 3 \bmod 7$, $297 \equiv 0 \bmod 11$. $\overline{2}$ has order 4 in $U(\mathbb{Z}_5)$ and $\overline{3}$ has order 6 in $U(\mathbb{Z}_7)$. Thus, $297^{792} \equiv 2^{792} \bmod 5 \equiv 1 \bmod 5$ and
$297^{792} \equiv 3^{792} \bmod 7 \equiv 1 \bmod 7$. $297^{792} \equiv 0 \bmod 11$.

**8.** (i) $Hr_1 = \{r_1, s_3\}$, $Hr_2 = \{r_2, s_2\}$;

(ii) $r_1H = \{r_1, s_1\}$, $r_2H = \{r_2, s_2\}$;

(iii) $Hs_1 = \{s_1, r_3\}$, $Hr_1^{-1} = Hr_3 = \{r_3, s_1\} = Hs_1 = Hs_1^{-1}$;

(iv) $r_1K = \{r_1, s_3\}$.

## Chapter 11

**1.** $\text{orb}(2) = \{2, 1, 4\}$, $2 \leftrightarrow \text{send}_2(2) = \{e, s_1\}$, $1 \leftrightarrow \text{send}_2(1) = \{r_2, s_3\}$, $4 \leftrightarrow \text{send}_2(4) = \{r_1, s_2\}$.

**2.** $\text{fix}(e) = \{1, 2, 3, 4\}$, $\text{fix}(r_1) = \{3\} = \text{fix}(r_2)$, $\text{fix}(s_1) = \{2, 3\}$, $\text{fix}(s_2) = \{1, 3\}$, $\text{fix}(s_3) = \{4, 3\}$.Thus, $\sum_{g\in G} |\text{fix}(g)| = 12$.
$\text{stab}(1) = \{e, s_2\}$, $\text{stab}(2) = \{e, s_1\}$, $\text{stab}(3) = D_3$, $\text{stab}(4) = \{e, s_3\}$. Thus, $\sum_{x\in X} |\text{stab}(x)| = 12$. There are two orbits and $|D_3| = 6$.

**3.** $\text{fix}(\text{ref}_{\frac{\pi}{2}}) = \{(x, x) : x \in \mathbb{R}\}$ which is the line $y = x$. $\text{fix}(\text{rot}_\pi) = \{(0, 0)\}$.

**4.** $|\text{fix}(e)| = n^2$, $|\text{fix}(r_i)| = 0$ for $i = 1, 2, 3$, $|\text{fix}(s_2)| = |\text{fix}(s_4)| = n$, $|\text{fix}(s_1)| = |\text{fix}(s_3)| = 0$. The number of orbits is $\frac{n(n+2)}{8}$.

## Chapter 12

**1.** $\frac{28+4}{4} = 8$.

**2.** (i) (a) $\frac{n^4+3n^2}{4}$, (b) $\frac{6+(2\times 3)}{4} = 3$.

(ii) (a) $\frac{2^9+2^3+2^3+2^5}{4} = 140$, (b) $\frac{2^9+2^3+2^3+2^5+(2^6\times 4)}{8} = 102$.

**3.** There are 48 symmetries of which 24 are rotations.

**4.** (i) $|\text{orb(top face)}| = 2$ and $|\text{stab(top face)}| = 4$ so there are eight symmetries, four of which are rotations.

(ii) With the non-square rectangle as top face we have $|\text{orb(top face)}| = 4$ and $|\text{stab(top face)}| = 4$ so there are 16 symmetries, eight of which are rotations.

**5.** $\frac{15+(1\times 6)+(3\times 3)+0+(3\times 6)}{24} = 2$.

**6.** $\frac{6+0+(3\times 2)}{12} = 1$.

## Chapter 13

**1.** (i) $r_2, s_3$; (ii) $\text{ref}_{2\phi-\theta}, \text{rot}_\theta$.

**2.** $8 = 1 + 1 + 2 + 2 + 2$.

**3.** $0, \pi$.

**4.** (i) (4,3,2); (ii) (4,4,1); (iii) (2,1,1,1,1,1,1,1).

**5.** $\theta = \begin{pmatrix} 1 & 2 & 3 & 4 & 5 & 6 & 7 & 8 & 9 \\ 7 & 4 & 6 & 3 & 2 & 1 & 9 & 8 & 5 \end{pmatrix}$ (not unique).

**6.** $(1\ 2)(3\ 4)$, $(1\ 2)$ (not unique).

**7.** (i) $\text{cent}_{S_3}((1\ 2\ 3)) = \{\text{id}, (1\ 2\ 3), (1\ 3\ 2)\} =< (1\ 2\ 3) >$;
$\text{cent}_{S_3}((1\ 2)) = \{\text{id}, (1\ 2)\} =< (1\ 2) >$.

(ii) $\text{cent}_{D_4}(r_2) = D_4 \neq< r_2 >$; $\text{cent}_{D_4}(s_1) = \{e, s_1, s_3, r_2\} \neq< s_1 >$.

**8.** In both cases the centralizer is $G$.

**9.** Any element in $Z(S_3)$ must commute with both $(1\ 2\ 3)$ and $(1\ 2)$.

## Chapter 14

**1.** (i) $\text{orb}(P) = \{(0, 1, 1), (1, 0, 1), (1, 1, 0)\}$, $\text{stab}(P) = \{\text{id}, (1\ 2)\}$,
$\text{send}_P((0, 1, 1)) = \{(1\ 3), (1\ 2\ 3)\}$, 1; $\text{send}_P((1, 0, 1)) = \{(2\ 3), (1\ 3\ 2)\}$.

(ii) $\text{orb}(P) = \{(1, 1, 1)\}$, $\text{stab}(P) = S_3$.

**2.** (i) $g_3 = s_2$, 3 (ii) $g_3 = r_1$, 1 (iii) $(r_1, r_1, r_1)$, $(r_2, r_2, r_2)$, $(e, e, e)$.

**3.** (i) By Cauchy's Theorem, $G$ has elements of orders 3 and 2. These generate the required cyclic subgroups.

(ii) $ab$ has order 6 and so $G$ is cyclic generated by $ab$.

**4.** (i) Let $H =< a >$, $K =< b >$. Applying Theorem 2, $V = HK \cong C_2 \times C_2$.

(ii) $|V| = 4$. Apply Lagrange's Theorem.

**5.** Let $(g, h) \in D_4 \times C_2$. Then $g$ has order 1,2 or 4 and $h$ has order 1 or 2. Thus, $(g, h)$ has order 1,2, or 4. Hence, $D_4 \times C_2$ has no elements of order 8 whereas $D_8$ has one, namely $\text{rot}_{\frac{\pi}{4}}$.

**6.** Let $H =< \overline{2} >= \{\overline{1}, \overline{2}, \overline{4}, \overline{8}\}$. Let $K =< \overline{11} >= \{\overline{1}, \overline{11}\}$. Theorem 2 shows that $U(\mathbb{Z}_{15}) = HK \cong C_4 \times C_2$.

## Chapter 15

**1.** (i) $\ker f = \{mn : m \in \mathbb{Z}\}$. Not injective.

(ii) $\ker f = \{2k\pi : k \in \mathbb{Z}\}$. Not injective.

(iii) $\ker f = \{0\}$. Injective.

(iv) $\ker f = \{\text{id}, \text{rot}_\pi\}$. Not injective.

**2.** (i) $\{e\}$. Faithful (ii) $\{e, s_1, s_3, r_2\}$. Not faithful.

**3.** (i) $r_3 H r_3^{-1} = \{e, s_3\}$; (ii) $s_3 H s_3^{-1} = \{e, s_1\} = H$; (iii) $r_3 H r_3^{-1} = \{e, r_2, s_4, s_2\} = H$.

**4.** (i) If $\text{ref}_\phi \in N$ then $\text{ref}_\theta = \text{rot}_{\frac{\theta-\phi}{2}} \text{ref}_\phi (\text{rot}_{\frac{\theta-\phi}{2}})^{-1} \in N$ for all $\theta$.

(ii) $\text{ref}_\phi \text{rot}_\theta \text{ref}_\phi = \text{rot}_{-\theta} \in< \text{rot}_\theta >$; $\text{rot}_\phi \text{rot}_\theta \text{rot}_{-\phi} = \text{rot}_\theta \in< \text{rot}_\theta >$.

**5.** Let $z \in Z(G)$. Then $gzg^{-1} = z \in Z(G)$ for all $g \in G$.

**6.** Since $Z(S_n) = \{e\}$ (see Example 3 of Chapter 13), $S_n$ has no normal subgroups of order 2. $A_n$ is a subgroup of index 2.

7. A normal subgroup is a union of conjugacy classes and its order divides that of the group. The class equation, $12 = 1 + 3 + 4 + 4$, tells us that the only possibilities in $A_4$ are {id}, {id, (1 2)(3 4), (1 3)(2 4), (1 4)(2 3)} and the whole group $A_4$. If $A_4$ had a subgroup of order 6 it would be normal since its index is 2. No such subgroup exists.

8. Let $\tau = (5\ 6)$. Then $\tau$ is odd and $\tau\alpha = \alpha\tau$. Apply Theorem 9(ii).

9. $\theta = \begin{pmatrix} 1 & 2 & 3 & 4 & 5 & 6 \\ 4 & 5 & 6 & 1 & 2 & 3 \end{pmatrix} = (1\ 4)(2\ 5)(3\ 6)$. Apply Theorem 9(ii).

## Chapter 16

1. $K = \{e, r_2, r_4\} \to$ id, $r_1K = \{r_1, r_3, r_5\} \to (1\ 3)(2\ 4)$, $s_1K = \{s_1, s_3, s_5\} \to$ $(1\ 4)(2\ 3)$, $s_2K = \{s_2, s_4, s_6\} \to (1\ 2)(3\ 4)$. This is Klein's 4-group.

2. $|A_4| = 12$, $|V| = 4$ so $|A_4/V| = 3$ is prime and $A_4/V$ is cyclic.

3. The order of each non-neutral element is 2 so $G/N$ is not cyclic.

4. (i) 2 (ii) $g^4N$ has order 3, $g^3N$ has order 2.

5. Let $f : GL_2(\mathbb{R}) \to (\mathbb{R}\backslash\{0\}, \times)$ be given by $f(A) = \det(A)$. Then $\ker f = SL_2(\mathbb{R})$. Since $f$ is surjective the result follows by the First Isomorphism Theorem.

6. The order of $Z(G)$ is 7,13 or 91. If $Z(G)$ has order 7 or 13 then $|G/Z(G)|$ is prime and so $G/Z(G)$ is cyclic. By Theorem 10, $G$ is abelian and so $Z(G) = G$. If $Z(G)$ has order 91, $G$ is clearly abelian.

7. $G = S_3$, $N = C_3 =< r_1 >$.

## Chapter 17

1. $a^3b = ba^2$, $\text{ref}_{\frac{6\pi}{5}} = \text{ref}_{\frac{-4\pi}{5}}$ and $ba^2b = a^3$, $\text{rot}_{\frac{-4\pi}{5}} = \text{rot}_{\frac{6\pi}{5}}$.

2. $Z(Q) = \{1, -1\}$, $\text{cent}_Q(i) = \{1, -1, i, -i\}$, $\text{conj}_Q(i) = \{i, -i\}$. The class equation is $1 + 1 + 2 + 2 + 2 = 8$.

3. (i) $a^2 = (1\ 9\ 7)(5\ 3\ 11)(12\ 4\ 6)(8\ 10\ 2)$, $a^3 = (1\ 3)(2\ 4)(5\ 7)(6\ 8)(9\ 11)(10\ 12)$, $a^4 =$ $(1\ 7\ 9)(5\ 11\ 3)(12\ 6\ 4)(8\ 2\ 10)$, $a^5 = (1\ 11\ 7\ 3\ 9\ 5)(12\ 2\ 6\ 10\ 4\ 8)$, $ba =$ $(1\ 6\ 3\ 8)(2\ 9\ 4\ 11)(5\ 10\ 7\ 12)$, $ba^2 = (1\ 10\ 3\ 12)(2\ 5\ 4\ 7)(6\ 9\ 8\ 11)$, $ba^3 =$ $(1\ 4\ 3\ 2)(5\ 8\ 7\ 6)(9\ 12\ 11\ 10)$, $ba^4 = (1\ 8\ 3\ 6)(2\ 11\ 4\ 9)(5\ 12\ 7\ 10)$, $ba^5 =$ $(1\ 12\ 3\ 10)(2\ 7\ 4\ 5)(6\ 11\ 8\ 9)$. For $0 \leqslant i \leqslant 5$ $ba^i$ has order 4; only $a^3$ has order 2.

   (ii) Let $X = \{t_1, \ldots, t_{12}\}$ be the set of the 12 terms in $p$. Then $G = \text{stab}(p)$ acts on $X$ and $\text{orb}(t_1) = X$. Let $g \in \text{stab}(t_1)$ then $g(1) = 1$, $g(6) = 6$ and so $g * t_{11} = t_{11}$ which gives $g(11) = 11$ and $g(10) = 10$. Hence, $g * t_7 = t_7$ and $g(7) = 7$, $g(4) = 4$. Thus, $g * t_3 = t_3$, $g(3) = 3$, $g(8) = 8$, $g * t_9 = t_9$, $g(9) = 9$, $g(12)$, $g * t_5 = t_5$ and $g(2) = 2$, $g(5) = 5$. We have shown that $g =$ id. The finite form of the Orbit-Stabilizer Theorem shows that $|\text{stab}(p)| = 12 \times 1 = 12$.

   (iii) This follows easily from (i).

**4.** $a^2$ and $a^4$ have order 3. Thus, $ab$ cannot have order 3 since $b \neq a, a^3$. Since $b \neq a^2$, $ab \neq a^3$ and so $ab$ cannot have order 2. $a$ and $a^5$ have order 6. Hence, $ab$ cannot have order 6 as $b \neq e$, $b \neq a^4$. The only possibility left is that $ab$ has order 4. Thus, $(ab)^2$ has order 2 and so $(ab)^2 = b^2$. Thus, $ab = b^2(ab)^{-1} = b^2b^{-1}a^{-1} = ba^5$.

# Glossary

| | | *Section* | *Page* |
|---|---|---|---|
| $A_n$ | alternating group | 7.5 | 74 |
| $C_n$ | cyclic group of order $n$ | 9.5 | 102 |
| $C_\infty$ | infinite cyclic group | 9.5 | 102 |
| $\mathbb{C}$ | complex numbers | 4.1 | 38 |
| $\text{cent}_G(a)$ | centralizer of $a$ in $G$ | 13.4 | 137 |
| $\text{conj}_G a$ | conjugacy class of $a$ in $G$ | 13.2 | 133 |
| $D_4$ | group of symmetries of a square | 1.1 | 1 |
| $D_n$ | dihedral group | 5.3 | 52 |
| $f^{-1}$ | inverse of the function $f$ | 2.2 | 14 |
| $\text{fix}(g)$ | fixed set of $g$ | 11.2 | 119 |
| $\langle g \rangle$ | cyclic subgroup generated by $g$ | 6.2 | 58 |
| $\|G\|$ | order of $G$ | 4.3 | 41 |
| $GL_n(\mathbb{R})$ | general linear group | 3.1 | 29 |
| h.c.f. | highest common factor | 4.1 | 38 |
| ker * | kernal of an action * | 15.2 | 151 |
| $\ker f$ | kernel of $f$ | 15.1 | 150 |
| l.c.m. | least common multiple | 4.1 | 38 |
| $\mathbb{N}$ | natural numbers | 4.1 | 38 |
| $O_2$ | orthogonal group | 1.2 | 7 |
| $\text{orb}(x)$ | orbit of $x$ | 7.2 | 71 |
| $\mathbb{Q}$ | rational numbers | 4.1 | 38 |
| $Q$ | quarternion group | 17.3 | 178 |
| $\mathbb{R}$ | real numbers | 2.2 | 14 |
| $\text{ref}_\phi$ | reflection | 1.1 | 1 |
| $\mathbb{R}^2$ | Euclidean plane | 2.2 | 14 |
| $\text{rot}_\phi$ | rotation | 1.1 | 1 |
| $S_4$ | permutations of $\{1, 2, 3, 4\}$ | 2.1 | 12 |
| $S_n$ | symmetric group | 4.4 | 41 |
| $S_X$ | all permutations of $X$ | 2.3 | 19 |
| $\text{send}_x(y)$ | set of elements that send $x$ to $y$ | 8.4 | 85 |
| $\text{sgn } f$ | sign of permutation $f$ | 7.5 | 74 |
| $\text{stab}(x)$ | stabilizer of $x$ | 7.3 | 72 |
| $U(\mathbb{Z}_n)$ | units of $\mathbb{Z}_n$ | 8.5 | 87 |
| $\mathbb{Z}$ | integers | 4.1 | 38 |
| $\mathbb{Z}_n$ | integers modulo $n$ | 8.5 | 87 |
| $Z(G)$ | centre of $G$ | 13.5 | 137 |

# Bibliography

There are many books on Group Theory so we have made no attempt to give a complete bibliography here. We have included details of any book referred to in the text as well as a few others we feel might be of interest. We have included two books that concentrate on the history of algebra and two biographies of individual mathematicians. Brief details of most of the other mathematicians mentioned can be found in a good biographical dictionary. We recommend the *Chambers Biographical Dictionary*, which is strong on mathematicians.


Allenby, R.B.J.T. *Linear Algebra,* Edward Arnold (in this series).

Allenby, R.B.J.T. *Rings, Fields and Groups,* 2nd ed., Edward Arnold, London, 1991.

Allenby, R.B.J.T. and Redfern, E.J. *Introduction to Number Theory with Computing,* Edward Arnold, London, 1989.

Armstrong, M.A. *Groups and Symmetry,* Springer-Verlag, New York, 1988.

Artin, M. *Algebra,* Prentice-Hall, New Jersey, 1991.

Belhoste, B. *Augustin-Louis Cauchy,* Springer-Verlag, New York 1991.

Birkhoff, G. and MacLane, S. *A Survey of Modern Algebra*, 3rd ed., Macmillan, New York, 1965.

Burnside, W. *Theory of Groups of Finite Order*, 2nd ed., Cambridge University Press (1911) (reprinted by Dover Publications Inc in 1955).

Hirst, K.E. *Numbers, Sequences, and Series*, Edward Arnold (in this series).

Kostrikin, A.I. and Shafarevich, I.R. (Eds) *Algebra 1, Encyclopaedia of Mathematical Sciences, vol 11*, Springer-Verlag, Berlin Heidelberg, 1990.

Martin, G.E. *Transformation Geometry.* Springer-Verlag, New York, 1982.

Novy, L. *Origins of Modern Algebra,* Noordhoff International, Leyden, 1973.

Ore, O. *Niels Henrik Abel,* Chelsea, New York, 1974.

van der Waerden, B.L. *A History of Algebra,* Springer-Verlag, Berlin Heidelberg, 1985.

# Index


Abel, N.H., 5, 187
Abelian group, 5
Action
  induced, 41
  of a group, 5, 68
  faithful, 152
Addition modulo $n$, 87
Additive notation, 41
Alternating
  group, 74
  polynomial, 74
Associative law, 5
Associativity, 5, 15, 41
Automorphism, 192

Bijection, 17
Bijective function, 17
Binary operation, 40
Burnside, W., 122, 187

Cancellation laws, 45
Carmichael number, 115
Cartesian product, 46
Cauchy's Theorem, 143
Cauchy, A.L., 142, 187
Cayley table, 4
Cayley's Theorem, 102
Cayley, A., 102, 187
Central element, 137
Centralizer, 137
Centre, 137
Character, 190
Class equation, 135
Closure, 4, 41
Codomain of a function, 14
Comute, 5, 41
Complex number, 38
Composite of two functions, 15
Composition of functions, 14
Conjugacy class, 133
Conjugate of an element, 133
  of a subgroup, 152
Coprime, 39
Coset, 6
  left, 106
  right, 114
Cube, symmetries of, 127
Cyclic group, 57
  subgroup, 58
Cycle, 20
Cycle decomposition, 22
Cycle type, 135

De Moivre's Theorem, 39
Determinant of a matrix, 31
Dicyclic group, 183
Dihedral group, 1, 42
Direct product, 46
Disjoint cycles, 21
Division ring, 189
Dodecahedron, 132
Domain of a function, 14

Enlargement, 33
Equivalence
  modulo $H$, 108
  class, 81
  modulo $n$, 81
  relation, 81
Euclidean plane, 14
Euler, L., 186

Factor group, 166
Faithful action, 152
Fermat's Little Theorem, 113
Fermat, P. de, 186
Field, 189
15-puzzle, 26
First Isomorphism Theorem, 170
Fixed set, 119
Flow, 192
Frieze patterns, 180
Function, 14
Fundamental theorem of arithmetic, 39

Galois, E., 187
Gauss, C.F., 186
General linear group, 32, 42
Generator of a cyclic group, 57
Generators of a group, 188
Glide, 188
Group 1, 41
  abelian, 41
  alternating, 74
  axioms, 41
  commutative, 41
  cyclic, 57
  definition, 41
  dicyclic, 181
  dihedral, 1, 42, 54
  factor, 166
  free, 189
  Galois, 189
  general linear, 32, 42
  linear, 190
  of *n*th roots of unity, 51
  of symmetries of a figure, 53
  of symmetries of a square, 1
  of units of a ring, 189
  one-parameter, 190
  orthogonal, 42
  quarternion, 179
  rotation, 54, 128
  simple, 159
  special linear, 51
  special orthogonal, 52, 101
  symmetric, 12, 42
  direct product of, 46

Hamilton, W.R., 179, 187
Highest common factor (h.c.f.), 39
Homomorphism, 33, 94
  natural, 166

Icosahedron, 132
Identity
  element, 4, 41
  function, 15
  matrix, 30
  permutation, 12
Image, 52, 99
Inclusion, 97
Injection, 17
Injective function, 17
Integers, 38
Interlacing shuffle, 66
Inverse
  of an element, 5, 41
  of a function, 16
  of a matrix, 32
Inverses of products, 43
Invertible
  function, 18
  matrix, 31
Isomorphic, 94
Isomorphism, 194

Jordan, C., 186

Kernel
  of a homomorphism, 150
  of an action, 151
Klein's 4-group, 95
Klein, F., 95, 187

Lagrange, J.L. de, 186
Lagrange's Theorem, 110
Latin square property, 5, 45
Lie, S., 187
Linear transformation, 32
Least common multiple (l.c.m.), 39

Map, 14
Mapping, 14
Matrix, 29
  diagonal, 34
  identity, 30
  invertible, 31
  multiplication, 29
  orthogonal, 34
  square, 29
  symmetric, 34
  transpose of a
  upper triangular, 34
  zero, 36
Multiplication modulo *n*, 87

Neutral element, 4, 41
Non-trivial subgroup, 49
Normal subgroup, 154

Number
complex, 38
natural, 38
prime, 39
rational, 38

Octahedron, symmetries of, 129
Orbit, 6, 13, 71
Orbit-Counting Theorem, 121
Orbit-Stabilizer Theorem, 117
Order
of a group, 41
of an element, 59
Orthogonal
group, 41
matrix, 34

Partition, 81
Permutation, 5, 19
even, 76
odd, 76
sign of a, 76
Polar coordinates, 1
Polar form, 38
Polynomial
alternating, 74
orbit of a, 13
stabilizer of, 13
symmetric, 13
Projection, 33
Product of subsets, 145
Proper subgroup, 49

Quaternion, 179
Quaternion group, 179

Rectangle symmetries of, 42
Reflection, 2
Reflexive relation, 83
Relations, 82
equivalence, 83
reflexive, 83
symmetric, 83
transitive, 83
Relations of a group, 188
Representation, 190
Ring, 189
Rotation, 2
Rotation group, 54, 127
Ruffini, P., 186

Shear, 33
Simple group, 159
Special linear group, 51
Special orthogonal group, 51
Stabilizer, 5, 13, 72
Subgroup, 5, 49
cyclic, 51, 58
diagonal, 56
non-trivial, 49
normal, 154
of index 2, 156
proper, 49
trivial, 49
Subgroup criterion, 49
Surjection, 17
Surjective function, 17
Sylow, L., 188
Sylow Theorems, 188
Symmetric
group, 12, 42
matrix, 34
polynomial, 14
relation, 82
Symmetries, 1, 52

Tetrahedron, 128
Transitive relation, 83
Translation, 188
Transpose of a matrix, 29
Transposition, 24
adjacent, 25
Trivial subgroup, 49

Upper triangular matrix, 34

von Dyck, W., 187

Wallpaper patterns, 188
Weber, H., 187
Wilson's Theorem, 113